APPLIED CODEOLOGY

NAVIGATING THE 2020 *NEC*®

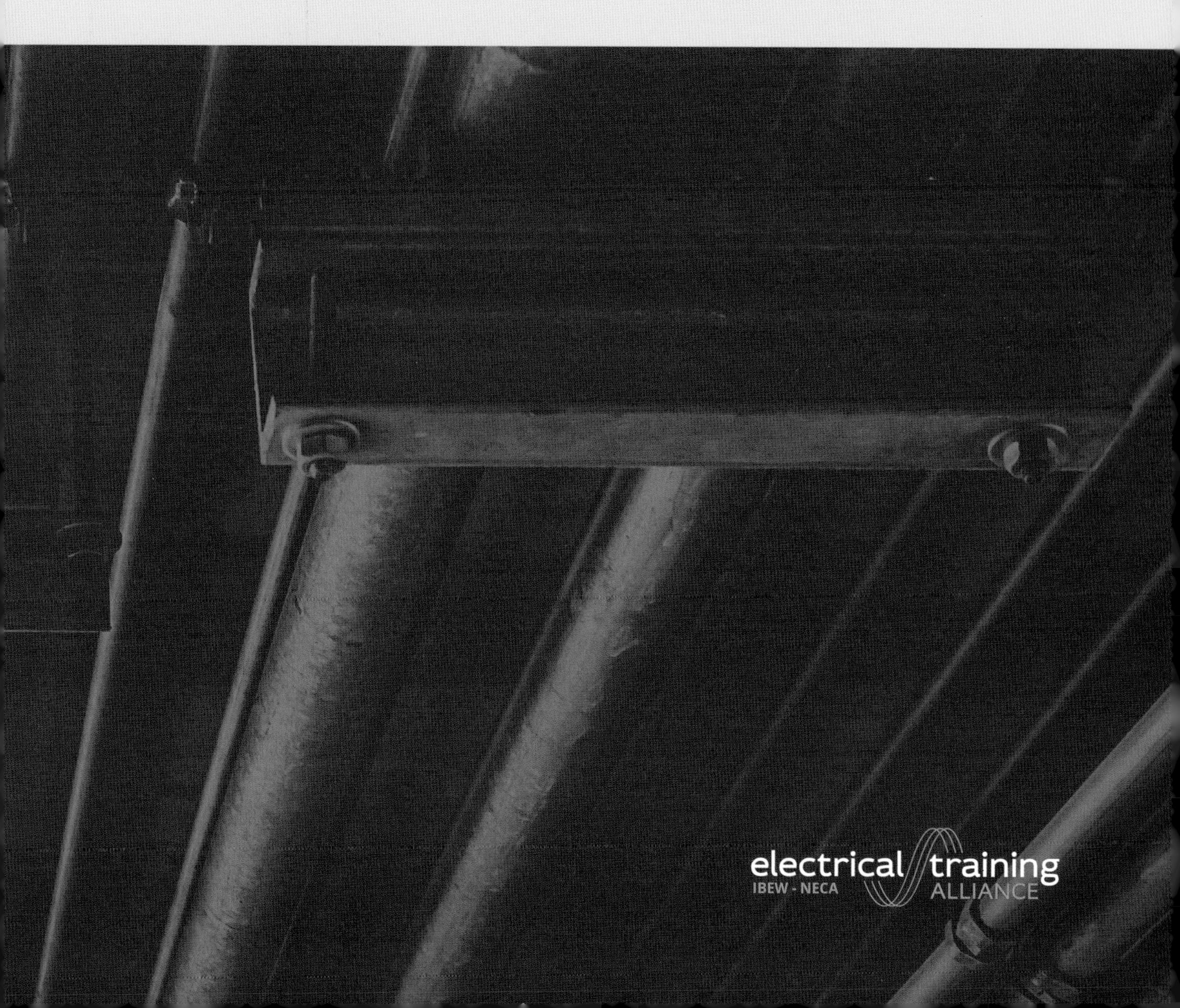

Applied Codeology is intended to be an educational resource for the user and contains procedures commonly practiced in industry and the trade. Specific procedures vary with each task and must be performed by a qualified person. For maximum safety, always refer to specific manufacturer recommendations, insurance regulations, specific job site and plant procedures, applicable federal, state, and local regulations, and any authority having jurisdiction. The *electrical training ALLIANCE* assumes no responsibility or liability in connection with this material or its use by any individual or organization.

1 2 3 4 5 6 7 8 9 – 20 – 9 8 7 6 5 4 3 2 1

Printed in the United States of America

M66002

Contents

Contents

Contents

Features

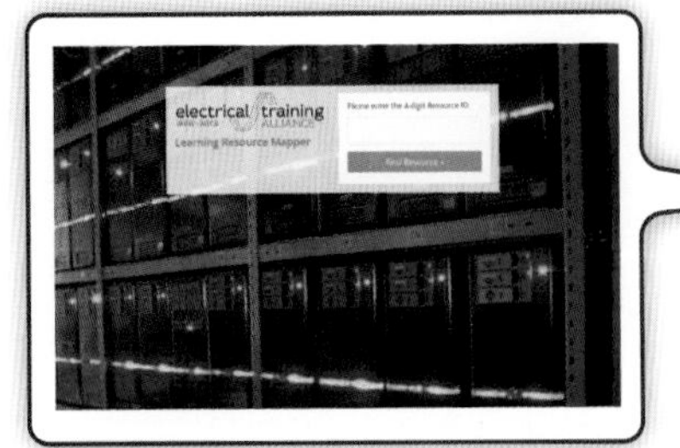

For additional information related to QR Codes, visit qr.njatcdb.org Item #1079

Quick Response Codes (QR Codes) create a link between the textbook and the Internet. They can be scanned using Smartphone applications to obtain additional information online. (To access the information without using a Smartphone, visit qr.njatcdb.org and enter the referenced Item #.)

Figures, including photographs and artwork, clearly illustrate concepts from the text.

glance through **Article 422** shows that this term means much more. The definition for *appliance* from **Article 100** is:

> **Appliance:** Utilization equipment, generally other than industrial, that is normally built in standardized sizes or types and is installed or connected as a unit to perform one or more function such as clothes washing, air-conditioning, food mixing, deep frying, and so forth.

Article 422 covers electrical appliances regardless of the occupancy in which they are used. However, while the definition of *appliance* technically includes air-conditioning equipment, this equipment is covered separately in **Article 440**, Air-Conditioning and Refrigerating Equipment. It is also important to consider that many electrical appliances are motor-operated, in which case the requirements in **Article 430** must be followed.

Section 422.5 lists those appliances that require GFCI protection. This provides a convenient way to determine whether an appliance being installed requires GFCI protection by one of the means covered in **422.5(B)**. Note that this does not negate other GFCI requirements included in the *NEC*, and **Section 210.8** is specifically referenced here.

Below is a list of appliances that have specific requirements included in **Article 422**:

- Central heating equipment
- Storage-type water heaters
- Central vacuum outlets
- Garbage disposers
- Built-in dishwashers
- Built-in trash compactors
- Wall-mounted ovens
- Counter-mounted cooking uni[illegible]
- Range hoods
- Microwave oven / range [illegible]d combo units
- Ranges

There are other appliances [illegible]h requirements in other articles, [illegible]ch as ceiling fan supports in **Ar[illegible] 314**, which **Article 422** directs t[illegible]er to.

Figure 4-4 Panelboard

Figure 4-4. Panelboards are typically used for branch circuits to supply wiring devices such as receptacles and luminaires.

Any piece of equipment will require maintenance or eventual replacement. **Part III** of **Article 422** contains the disconnect requirements for appliances. In many cases, smaller appliances shall be permitted to use the cord and plug to disconnect. In others, the disconnecting means must generally be either in sight or capable of being locked in the open position.

> **In Sight From (Within Sight From, Within Sight).** Where this *Code* specifies that one equipment shall be "in sight from," "within sight from," or "within sight of," and so forth, another equipment, the specified equipment is to be visible and not more than 15 m (50 ft) distant from the other.

Articles 424 to 427, Heating For Spaces, Processes, Pipelines, and Snow Melting Equipment

These four articles cover types of hea[illegible] ing that, with the exception of space heating in **Article 424**, are primarily

Whenever a circuit is installed for survivability of critical circuits, **Article 728, Fire-Resistive Cable Systems**, comes into play. Electrical circuit protective systems must be installed exactly as listed. Circuit integrity cables typically need to be supported every two feet, as that is how they are tested to achieve their listing. Where cables are required to be in a raceway, the raceway is typically supported every four to six feet, depending on the system. Connectors, couplings, vertical or horizontal orientations, and even the brand of raceway must all align with the listing requirements.

Perhaps the most important section of **Article 728** from an installer's point of view is **728.120**. Per this section, fire resistive cables are required to be marked with the system identifier. This identifier can be used to identify the proper way to install the system within the requirements of the listing.

For additional information, visit qr.njatcdb.org Item #5517

Article 760, Fire Alarm Systems

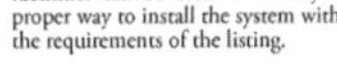

Fact

Most people do not have a fire alarm system in their home. The large majority of homes have what are called single-station or multi-station smoke alarms. These devices do not require a fire alarm panel to operate and are not within the scope of **Article 760**.

Many of the requirements i[illegible] **Article 760** are also included in **Art[illegible]cle 725**, since both articles are conce[illegible]ed with systems that are typically lo[illegible]er-voltage systems. As such, **Art[illegible]le 760** shares a parallel numbering [illegible]ructure with **Article 725**.

The article on fire alarm [illegible]tems is broken down into four parts[illegible] **Part I** is general and applies to all fire [illegible]rm systems. **Section 760.3** lists wh[illegible]ections in **Article 300** apply to fire [illegible]rm systems. **Section 760.24** inclu[illegible] the requirements for installation. [illegible]te that while only **300.4(D)** applies [illegible]**25.24**, due to the critical nature o[illegible] alarm

Pull stations are located within five feet from the exit of a floor per NFPA 72. Covers discourage malicious activation of the manual alarm device.

systems, all of **Section 300.4** applies within **760.24**.

While it is common for junction boxes containing fire alarm conductors to be identified with red paint in order to comply with **760.30**, the color red is not required. However, the box must be identified in some way to prevent unintentional signals on the fire alarm system.

In most jurisdictions, fire alarm systems must also be installed according to *NFPA 72: The National Fire Alarm and Signaling Code.*

*Pull stations must be located where they are conspicuous, unobstructed, and accessible. A blocked pull station does not comply with these requirements from NFPA 72, **Chapter 17**.*

***Code* Excerpts** are "ripped" from *NFPA 70®* or other sources.

Headers and **Subheaders** organize information within the chapter.

Facts offer additional information related to the *Codeology* method.

Features

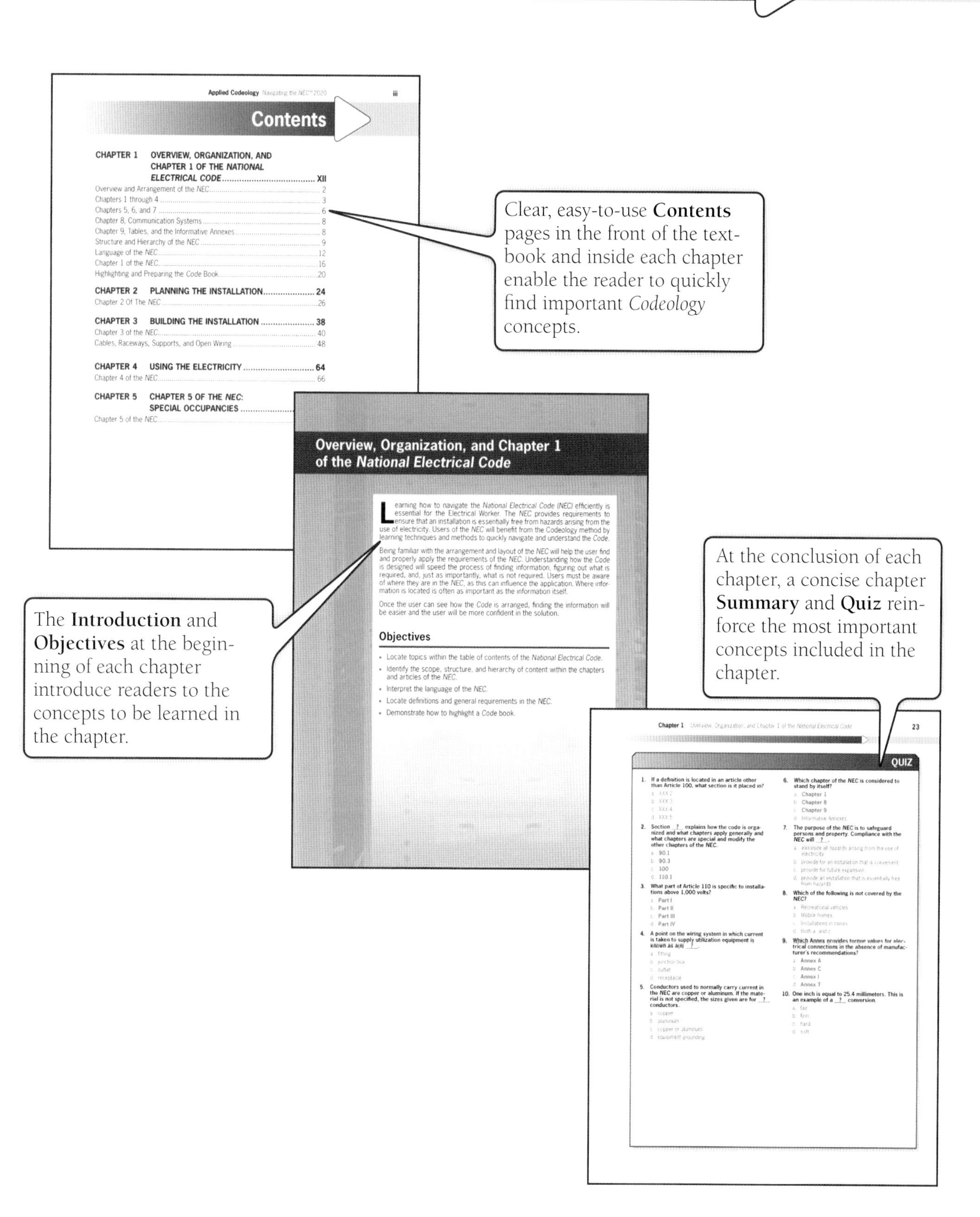
Clear, easy-to-use **Contents** pages in the front of the textbook and inside each chapter enable the reader to quickly find important *Codeology* concepts.
The **Introduction** and **Objectives** at the beginning of each chapter introduce readers to the concepts to be learned in the chapter.
At the conclusion of each chapter, a concise chapter **Summary** and **Quiz** reinforce the most important concepts included in the chapter.
Contents
CHAPTER 1 OVERVIEW, ORGANIZATION, AND CHAPTER 1 OF THE NATIONAL ELECTRICAL CODE XII
CHAPTER 2 PLANNING THE INSTALLATION 24
CHAPTER 3 BUILDING THE INSTALLATION 38
CHAPTER 4 USING THE ELECTRICITY 64
CHAPTER 5 CHAPTER 5 OF THE NEC: SPECIAL OCCUPANCIES
Overview, Organization, and Chapter 1 of the National Electrical Code
Objectives
QUIZ

Introduction

Apprentices, Journeymen Electrical Workers, electrical engineers, electrical maintenance workers, electrical inspectors, and numerous other Electrical Workers use *NFPA 70: National Electrical Code (NEC)* on a daily basis. Their livelihoods depend upon their ability to properly design, install, inspect, and maintain electrical systems in accordance with the *NEC*. The *NEC* can be overwhelming to new apprentices in the electrical industry. In fact, some Electrical Workers complete an apprenticeship or an electrical engineering degree without confidence in their ability to quickly find references within the *NEC*. This can be very frustrating since they must use the *NEC* on a daily basis to complete their work. Add to this the fact that the *NEC* changes every three years, and this frustration continues to grow.

An experienced tradesperson can differentiate between members of different trades just by looking at the tools they carry. Electrical Workers use hand tools, the most identifiable being side-cutting pliers. To an apprentice, a pair of side-cutters seems clumsy, but necessary. After a few years of field experience, those same side-cutters seem to become an extension of the hand, capable of many tasks, all of which can be performed quickly and efficiently. The *NEC* is another "tool" that Electrical Workers use on a daily basis. As with any other tool, the Electrical Worker must learn to use the *NEC* properly in order to successfully complete the task required. Only when an Electrical Worker understands how to use the *NEC* will he or she will have a complete set of tools and be able to efficiently complete the tasks required.

About This Book

The *electrical training ALLIANCE* textbook *Applied Codeology: Navigating the NEC® 2020* is designed to help the student understand how to use the *National Electrical Code (NEC)*. The term "*Codeology*" is derived from two words: "code," referring to the *National Electrical Code*, and "ology," meaning an academic field of study. Thus, "Applied Codeology" is the academic study of the *NEC*, both in the classroom and on the job.

The original concept of Codeology was developed in the 1960s by Ronald O'Riley, a prominent author of electrical training materials, who used it in his training program. Over the years, the *electrical training ALLIANCE* secured the rights to publish the book, and it has been a fundamental part of the IBEW apprenticeship training program for over four decades.

In order to be able to find information in the *NEC* efficiently, the user has to have an idea of what is in the *NEC* in the first place, and know how the *NEC* is generally organized. To better achieve this objective, this new edition of *Applied Codeology* has a vastly different format compared to previous editions. The first nine chapters of this textbook mirror the *NEC* on a chapter-by-chapter basis. Each of the first nine chapters of *Applied Codeology* summarize the articles in the corresponding chapter of the *NEC*. Once the reader has a basic familiarity with the *NEC* as afforded by the chapters in this textbook, the "Codeology" method is introduced in Chapter 10. This method, combined with a basic knowledge of the *NEC*, will enable a person to find the information he or she needs more efficiently.

Using the Codeology method, students will discover and describe the fundamental structure of the *NEC*. In addition, *Applied Codeology* will introduce methods and strategies that the user can apply immediately to improve the skills needed to find information in the parts, sections, and subdivisions used in the *Code*.

The *Code* is written in a very organized and methodical way. With an understanding of the organizational structure of the *NEC*, the *Code* user can efficiently and confidently maneuver through the articles and sections easily, finding the information needed to answer a question or complete a task. The objective of *Applied Codeology* is to provide the user with basic skills in understanding how the *Code* book is organized and to help students and users build confidence in their ability to quickly find pertinent information within the *NEC*. In a classroom setting, the content within the *Applied Codeology* textbook may lead to discussion of specific *Code* rules. Keep in mind that the intent of Codeology is to be a tool used to develop confidence in finding information in the *NEC*.

Acknowledgements

Special thanks to those who shared their expertise and encouragement during the writing of this book. The patience of the editorial staff of the *electrical training ALLIANCE*, whose efforts trying to get this electrician to understand grammar greatly contributed to the readability of this book, is exemplary. The other subject matter experts, Todd Dunne, Dave Sanchez, and Trevor Turek, each brought a different point of view of the chapters that the reader can be thankful for. The instructional staff at the *NECA* IBEW Training Center in Portland, Oregon, notably Topher Edwards, were a great encouragement with ideas and illustration help.

When I started as an apprentice, I had no idea about the great opportunities that would present themselves to me in this field. I owe a debt of gratitude to all the Journeyman electricians who taught a very green apprentice the ins and outs of the trade and how to pass on that tradition. While I became a Journeyman, I consider myself to be in my 29th year of apprenticeship—always learning.

This book would not be in your hands without the help of my wonderful wife of 30 years and my children, who shared my time with the industry to see the completion of this project. Thank you.

John McCamish

11/1/19

QR Credits

National Fire Protection Association (NFPA)
U.S. Department of Labor
Test and Measurement Tips: An EE Online World Resource
nVent
Google Maps
UL Product IQ

About the Author

John McCamish has been involved in the electrical industry for 30 years. After graduating from the University of Oregon, he pursued a career in the electrical industry, first as an apprentice, then as a Journeyman, and then as a foreman. After working for over 10 years in the field, he has instructed apprentice and Journey-level Electrical Workers in various aspects of the trade, including fire alarm, instrumentation, code and calculations, and theory, among others.

McCamish currently serves his community as a member of a statewide specialty code committee and serves nationally on the Fundamentals Technical Committee for *NFPA 72: The National Fire Alarm and Signaling Code* and on Code-Making Panel 2 of the *National Electrical Code*.

Overview, Organization, and Chapter 1 of the *National Electrical Code*

Learning how to navigate the *National Electrical Code (NEC)* efficiently is essential for the Electrical Worker. The *NEC* provides requirements to ensure that an installation is essentially free from hazards arising from the use of electricity. Users of the *NEC* will benefit from the Codeology method by learning techniques and methods to quickly navigate and understand the *Code*.

Being familiar with the arrangement and layout of the *NEC* will help the user find and properly apply the requirements of the *NEC*. Understanding how the *Code* is designed will speed the process of finding information, figuring out what is required and, just as importantly, what is not required. Users must be aware of where they are in the *NEC*, as this can influence the application. Where information is located is often as important as the information itself.

Once the user can see how the *Code* is arranged, finding the information will be easier and the user will be more confident in the solution.

Objectives

» Locate topics within the table of contents of the *National Electrical Code*.

» Identify the scope, structure, and hierarchy of content within the chapters and articles of the *NEC*.

» Interpret the language of the *NEC*.

» Locate definitions and general requirements in the *NEC*.

» Demonstrate how to highlight a *Code* book.

Table of Contents

OVERVIEW AND ARRANGEMENT OF THE *NEC*

The *NEC* is divided into ten separate divisions, starting with **Article 90** and ending with the Informative Annexes. Following **Article 90**, there are eight chapters with their own articles. The ninth chapter contains tables referenced elsewhere in the *NEC*, and following that chapter are the Annexes.

Table of Contents

The table of contents is a good place to start to become familiar with the layout of the *National Electrical Code*. The table of contents will also prove to be useful in finding answers in the *NEC*.

In the table of contents, each chapter is followed by a list of the articles within that chapter. Each of the articles is assigned a number, the first digit of which is the same as the chapter. For example, **Article 230, Services**, is in **Chapter 2**. Most articles, unless they are very brief, are broken down into parts to help organize the material within. The individual parts of each article are listed after the article in the table of contents.

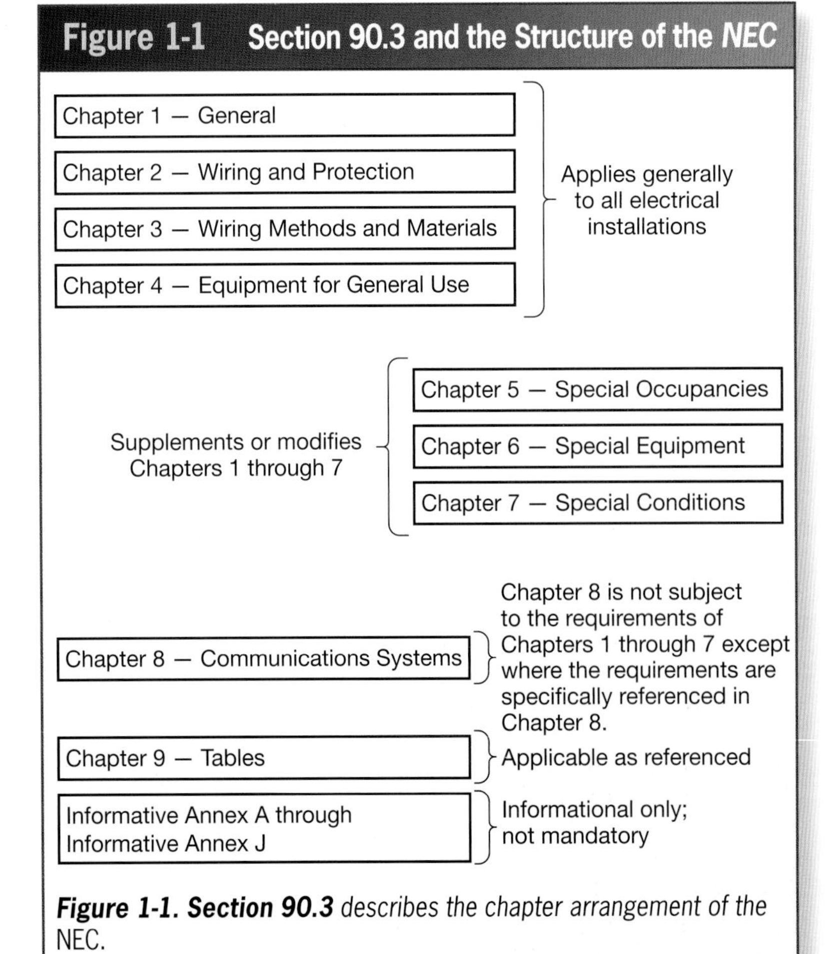

***Figure 1-1. Section 90.3** describes the chapter arrangement of the* NEC.

Article 90

Article 90 is the first article in the *NEC*, and is titled "Introduction." When specifying a section of an article, the article's number is followed by a period and then the number denoting the section of the article. **Article 90** has nine sections, numbered **90.1** through **90.9**.

Since **Article 90** is the introduction, it contains important information that affects the rest of the *NEC*. **Sections 90.1** and **90.2** spell out the purpose and the scope of the *NEC*. Each article in the *NEC* starts with the scope of the article. Each article will generally repeat this pattern, with scope of the article followed by more specific applications. Notice that the *NEC* as a whole starts with the scope of the entire document and becomes more specific with each chapter. This knowledge can be useful when searching for information in the *NEC* and within each article.

The scope of the *NEC* does not include every possible electrical installation, as is specified in **Section 90.2**. Ships, automotive vehicles, and installations in mines are among several applications or locations not included in the scope of the *NEC*.

Section 90.3 lays out the framework to use when applying the requirements in the *NEC*. This framework originally appeared in the 1937 edition of the *NEC* and has remained largely unchanged since that time. **Chapters 1** through **4** apply generally to all electrical installations. These chapters, therefore, have high priority in the Electrical Worker's *Code* "toolbox" since they have very far-reaching applications. The information in these chapters is used in the layout, planning, and installation of all electrical work. **Chapters 5, 6,** and 7 are for special occupancies, special equipment, and special conditions respectively. These chapters may modify or supplement **Chapters 1** through 7.

For example, a general lighting application would require the use of **Article 410, Luminaires, Lampholders, and Lamps.** The wiring method would be governed by **Chapter 3, Wiring Methods and Materials**. If the lighting was in an oil refinery, where hazardous conditions such as flammable liquids could be present, requirements in **Chapter 5** relating to hazardous locations would supplement the general requirements of the wiring and lighting.

Chapter 8 of the *NEC* covers communications. This chapter is a standalone chapter. Nothing in the previous seven chapters applies to **Chapter 8**, unless **Chapter 8** states otherwise. **Chapter 9** of the *NEC* contains tables that are applicable as referenced throughout the *NEC*. The Annexes have material that is not enforceable, but is used for informational purposes. **Figure 90.3** in the *NEC* shows the relationship of the various chapters. **See Figure 1-1.**

CHAPTERS 1 THROUGH 4

Chapters 1 through **4** form the four corners of the foundation on which the *NEC* is built. All electrical installations will depend on some part of these four chapters.

Chapter 1, General

Chapter 1 contains two articles. **Article 100, Definitions**, is important for anyone involved in the electrical industry. Conflicts regarding the right way to perform a task are often won or lost over the use and understanding of definitions.

Article 110 covers the general requirements for installations such as working spaces and terminations. This article applies to all chapters of the *NEC*.

Chapter 2, Wiring and Protection

Chapter 2 has nine articles; seven cover wiring and two cover protection. The articles in **Chapter 2** that deal with wiring also cover the identification of grounded conductors. Wiring from the service to feeders to the branch circuits is covered in separate articles. Proper load calculation methods and wire sizing are also covered in the first half of **Chapter 2**.

Article	Title
200	**Use and Identification of Grounded Conductors**
210	**Branch Circuits**
215	**Feeders**
220	**Branch-Circuit, Feeder, and Service Load Calculations**
225	**Outside Branch Circuits and Feeders**
230	**Services**

Article 250, Grounding and Bonding, covers the sizing and installation of some conductors. This article is included with the "protection" articles of **Chapter 2** because the proper grounding of a system is fundamental to the protection of the equipment, the property, and the persons using the electrical system. The first two articles of the protection section discuss the requirements for devices that protect against overcurrents and overvoltages in a system.

Article	Title
240	**Overcurrent Protection**
242	**Overvoltage Protection**
250	**Grounding and Bonding**

Chapter 3, Wiring Methods and Materials

Chapter 3 provides information on all permitted methods and materials needed to supply and install an electrical installation. This chapter details requirements for installation, from the service point to the termination at the last outlet in the electrical distribution system. The articles in **Chapter 3** provide general information for all wiring methods and all types of cable assemblies, raceways,

For additional information, visit qr.njatcdb.org Item #1562

cabinets, cutout boxes, meter socket enclosures, boxes, conduit bodies, fittings, and more.

Chapter 3 can be divided into five general topical categories. The first category addresses the general rules for all wiring methods, which include the general installation requirements of **Article 300** and requirements for conductors in **Article 310** and **Article 311**. The second category covers cabinets, enclosures, and boxes in **Article 312** and **Article 314**. The third category discusses the various types of cable assembles in **Article 320** through **Article 340**. Notice that the cable assemblies are addressed in alphabetical order. The fourth category covers raceways in **Article 342** through **Article 390**, starting with metal conduits, both rigid and flexible, followed by nonmetallic conduits and tubing. The methods covered in **Article 342** through **Article 362** tend to be wiring methods with a circular cross-section, while the methods discussed in **Article 366** through **Article 390** are mostly wiring methods with a rectangular cross section. Finally, the fifth category addresses open wiring or support methods in **Article 392** through **Article 399**.

Chapter 3 consists of 47 Articles in total:

Articles 300 and 310 cover general wiring requirements

300 General Requirements for Wiring Methods and Materials

310 Conductors for General Wiring

311 Medium Voltage Conductors and Cable

Articles 312 and 314 cover cabinets and enclosures of different kinds

312 Cabinets, Cutout Boxes, and Meter Socket Enclosures

314 Outlet, Device, Pull, and Junction Boxes; Conduit Bodies; Fittings; and Handhole Enclosures

Articles 320 to 340 cover cable assemblies

320 Armored Cable: Type AC

322 Flat Cable Assemblies: Type FC

324 Flat Conductor Cable: Type FCC

326 Integrated Gas Spacer Cable: Type IGS

330 Metal-Clad Cable: Type MC

332 Mineral-Insulated, Metal-Sheathed Cable: Type MI

334 Nonmetallic-Sheathed Cable: Types NM and NMC

336 Power and Control Tray Cable: Type TC

337 Type P Cable

338 Service-Entrance Cable: Types SE and USE

340 Underground Feeder and Branch-Circuit Cable: Type UF

Articles 342 to 362 cover raceways that have a round cross-section

342 Intermediate Metal Conduit: Type IMC

344 Rigid Metal Conduit: Type RMC

348 Flexible Metal Conduit: Type FMC

350 Liquidtight Flexible Metal Conduit: Type LFMC

352 Rigid Polyvinyl Chloride Conduit: Type PVC

353 High Density Polyethylene Conduit: Type HDPE Conduit

354 Nonmetallic Underground Conduit with Conductors: Type NUCC

355 Reinforced Thermosetting Resin Conduit: Type RTRC

356	**Liquidtight Flexible Nonmetallic Conduit: Type LFNC**
358	**Electrical Metallic Tubing: Type EMT**
360	**Flexible Metallic Tubing: Type FMT**
362	**Electrical Nonmetallic Tubing: Type ENT**

Articles 366 to 390 cover raceways that have a rectangular cross-section

366	**Auxiliary Gutters**
368	**Busways**
370	**Cablebus**
372	**Cellular Concrete Floor Raceways**
374	**Cellular Metal Floor Raceways**
376	**Metal Wireways**
378	**Nonmetallic Wireways**
380	**Multioutlet Assembly**
382	**Nonmetallic Extensions**
384	**Strut-Type Channel Raceway**
386	**Surface Metal Raceways**
388	**Surface Nonmetallic Raceways**
390	**Underfloor Raceways**

Articles 392 to 399 do not cover raceways, but cover support systems and open wiring

392	**Cable Trays**
393	**Low-Voltage Suspended Ceiling Power Distribution Systems**
394	**Concealed Knob-and-Tube Wiring**
396	**Messenger-Supported Wiring**
398	**Open Wiring on Insulators**
399	**Outdoor Overhead Conductors over 1000 Volts**

Chapter 4, Equipment for General Use

The fourth chapter of the *NEC* is about equipment that is generally used; more specialized equipment is found in **Chapter 6**. The equipment in **Chapter 4** uses electricity, controls electricity, or, in the case of transformers and generators, changes the voltage or provides a voltage source.

Article	Title
400	**Flexible Cords and Flexible Cables**
402	**Fixture Wires**
404	**Switches**
406	**Receptacles, Cord Connectors, and Attachment Plugs (Caps)**
408	**Switchboards, Switchgear, and Panelboards**
409	**Industrial Control Panels**
410	**Luminaires, Lampholders, and Lamps**
411	**Low-Voltage Lighting**
422	**Appliances**
424	**Fixed Electric Space-Heating Equipment**
425	**Fixed Resistance and Electrode Industrial Process Heating Equipment**
426	**Fixed Outdoor Electric De-icing and Snow-Melting Equipment**
427	**Fixed Electric Heating Equipment for Pipelines and Vessels**
430	**Motors, Motor Circuits, and Controllers**
440	**Air-Conditioning and Refrigerating Equipment**
445	**Generators**
450	**Transformers and Transformer Vaults (Including Secondary Ties)**

455	**Phase Converters**
460	**Capacitors**
470	**Resistors and Reactors**
480	**Storage Batteries**
490	**Equipment Over 1000 Volts, Nominal**

CHAPTERS 5, 6, AND 7

Chapters 5, 6, and 7 address special occupancies, special equipment, and special conditions respectively. These chapters modify or amend the general **Chapters 1** through **4** and the special **Chapters 5** through 7.

Chapter 5, Special Occupancies

Chapter 5 consists of 27 articles that offer supplemental rules for electrical installations in special occupancies. This includes locations that are hazardous due to processes that occur at those locations, health care facilities, aircraft hangers, carnivals, and agricultural buildings.

Article	Title
500	**Hazardous (Classified) Locations, Classes I, II, and III, Divisions 1 and 2**
501	**Class I Locations**
502	**Class II Locations**
503	**Class III Locations**
504	**Intrinsically Safe Systems**
505	**Zone 0, 1, and 2 Locations**
506	**Zone 20, 21, and 22 Locations for Combustible Dusts or Ignitable Fibers/Flyings**
510	**Hazardous (Classified) Locations - Specific**
511	**Commercial Garages, Repair and Storage**
513	**Aircraft Hangars**
514	**Motor Fuel Dispensing Facilities**
515	**Bulk Storage Plants**
516	**Spray Application, Dipping, Coating, and Printing Processes Using Flammable or Combustible Materials**
517	**Health Care Facilities**
518	**Assembly Occupancies**
520	**Theatres, Audience Areas of Motion Picture and Television Studios, Performance Areas, and Similar Locations**
522	**Control Systems for Permanent Amusement Attractions**
525	**Carnivals, Circuses, Fairs, and Similar Events**
530	**Motion Picture and Television Studios and Similar Locations**
540	**Motion Picture Projection Rooms**
545	**Manufactured Buildings and Relocatable Structures**
547	**Agricultural Buildings**
550	**Mobile Homes, Manufactured Homes, and Mobile Home Parks**
551	**Recreational Vehicles and Recreational Vehicle Parks**
552	**Park Trailers**
555	**Marinas, Boatyards, Floating Buildings, and Commercial and Noncommercial Docking Facilities**
590	**Temporary Installations**

Chapter 6, Special Equipment

The scope of **Chapter 6** is modifications and/or supplemental information for electrical installations containing special equipment. This chapter includes "equipment-specific" information supplementing or modifying the first seven chapters in regards to special equipment such as electric signs, welders, x-ray equipment, swimming pools, solar photovoltaic systems, fuel cells, and fire pumps. **Chapter 6** consists of 27 articles in total.

Article	Title
600	**Electric Signs and Outline Lighting**
604	**Manufactured Wiring Systems**
605	**Office Furnishings**
610	**Cranes and Hoists**
620	**Elevators, Dumbwaiters, Escalators, Moving Walks, Platform Lifts, and Stairway Chairlifts**
625	**Electric Vehicle Power Transfer System**
626	**Electrified Truck Parking Spaces**
630	**Electric Welders**
640	**Audio Signal Processing, Amplification, and Reproduction Equipment**
645	**Information Technology Equipment**
646	**Modular Data Centers**
647	**Sensitive Electronic Equipment**
650	**Pipe Organs**
660	**X-Ray Equipment**
665	**Induction and Dielectric Heating Equipment**
668	**Electrolytic Cells**
669	**Electroplating**
670	**Industrial Machinery**
675	**Electrically Driven or Controlled Irrigation Machines**
680	**Swimming Pools, Fountains, and Similar Installations**
682	**Natural and Artificially Made Bodies of Water**
685	**Integrated Electrical Systems**
690	**Solar Photovoltaic (PV) Systems**
691	**Large-Scale Photovoltaic (PV) Electric Supply Stations**
692	**Fuel Cell Systems**
694	**Wind Electric Systems**
695	**Fire Pumps**

Chapter 7, Special Conditions

The scope of **Chapter 7** is modifications and/or supplemental information for electrical installations under special conditions. This chapter includes "condition-specific" information supplementing or modifying the first seven chapters for special conditions such as emergency systems; legally required standby systems; Class 1, 2, and 3 systems; and fire alarm systems. There are 15 articles in **Chapter 7**.

Article	Title
700	**Emergency Systems**
701	**Legally Required Standby Systems**
702	**Optional Standby Systems**
705	**Interconnected Electric Power Production Sources**
706	**Energy Storage Systems**
708	**Critical Operations Power Systems (COPS)**
710	**Stand-Alone Systems**
712	**Direct Current Microgrids**
720	**Circuits and Equipment Operating at Less Than 50 Volts**
725	**Class 1, Class 2, and Class 3 Remote-Control, Signaling, and Power-Limited Circuits**
727	**Instrumentation Tray Cable: Type ITC**
728	**Fire-Resistive Cable Systems**

750	**Energy Management Systems**
760	**Fire Alarm Systems**
770	**Optical Fiber Cables**

CHAPTER 8, COMMUNICATION SYSTEMS

Chapter 8 covers communications systems. These installations are not subject to the requirements of **Chapters 1** through 7 (per **Section 90.3**) except where the requirements are specifically referenced in the six articles of **Chapter 8**.

Article	Title
800	General Requirements for Communications Systems
805	Communications Circuits
810	Radio and Television Equipment
820	Community Antenna Television and Radio Distribution Systems
830	Network-Powered Broadband Communications Systems
840	Premises-Powered Broadband Communications Systems

CHAPTER 9, TABLES, AND THE INFORMATIVE ANNEXES

Chapter 9 contains tables that are referenced throughout the *NEC*. Per **Section 90.3**, **Chapter 9** tables are applicable as referenced and are part of the requirements of the *NEC*. The Informative Annexes, which follow **Chapter 9**, are not part of the requirements of the *NEC*, but are included for informational purposes.

Tables

Table 1	**Percent of Cross Section of Conduit and Tubing for Conductors and Cables**
Table 2	**Radius of Conduit and Tubing Bends**
Table 4	**Dimensions and Percent Area of Conduit and Tubing (Areas of Conduit or Tubing for the Combinations of Wires Permitted in Table 1, Chapter 9)**
Table 5	**Dimensions of Insulated Conductors and Fixture Wires**
Table 5A	**Compact Copper and Aluminum Building Wire Nominal Dimensions** and Areas**
Table 8	**Conductor Properties**
Table 9	**Alternating-Current Resistance and Reactance for 600-Volt Cables, 3-Phase, 60 Hz, 75°C (167°F) - Three Single Conductors in Conduit**
Table 10	**Conductor Stranding**
Table 11(A)	**Class 2 and Class 3 Alternating-Current Power Source Limitations**
Table 12(A)	**PLFA Alternating-Current Power Source Limitations**

Annexes

Informative Annex A	**Product Safety Standards**
Informative Annex B	**Application Information for Ampacity Calculation**
Informative Annex C	**Conduit, Tubing, and Cable Tray Fill Tables for Conductors and Fixture Wires of the Same Size**
Informative Annex D	**Examples**
Informative Annex E	**Types of Construction**

Informative Annex F	**Availability and Reliability for Critical Operations Power Systems; and Development and Implementation of Functional Performance Tests (FPTs) for Critical Operations Power Systems**
Informative Annex G	**Supervisory Control and Data Acquisition (SCADA)**
Informative Annex H	**Administration and Enforcement**
Informative Annex I	**Recommended Tightening Torque Tables from UL Standard 486A-486B**
Informative Annex J	**ADA Standards for Accessible Design**

STRUCTURE AND HIERARCHY OF THE *NEC*

The *NEC* is organized into chapters, with each chapter further divided into smaller articles and sections as needed. The hierarchy or the order in which the requirements are written is important to understanding and properly applying the *Code.*

Users of other National Fire Protection Agency (NFPA) documents are sometimes surprised to find that the *NEC* is organized differently than other NFPA documents, such as *NFPA 72: The National Fire Alarm and Signaling Code.* Most notably, the *NEC* is divided into chapters, then into articles, parts, and sections.

Chapters, Articles, and Parts

Chapters are divided into articles. Articles cover specific topics, such as branch circuits in **Article 210**, electrical metallic tubing in **Article 358**, or pipe organs in **Article 650**. If the article is large enough, it may warrant being divided into parts. Parts are used to organize an article into relevant subtopics. Parts are designated by roman numerals and titled according to the content they contain. **See Figure 1-2.**

> **Fact**
>
> The method and language used to organize the *NEC* can be found in the *NEC Style Manual.*

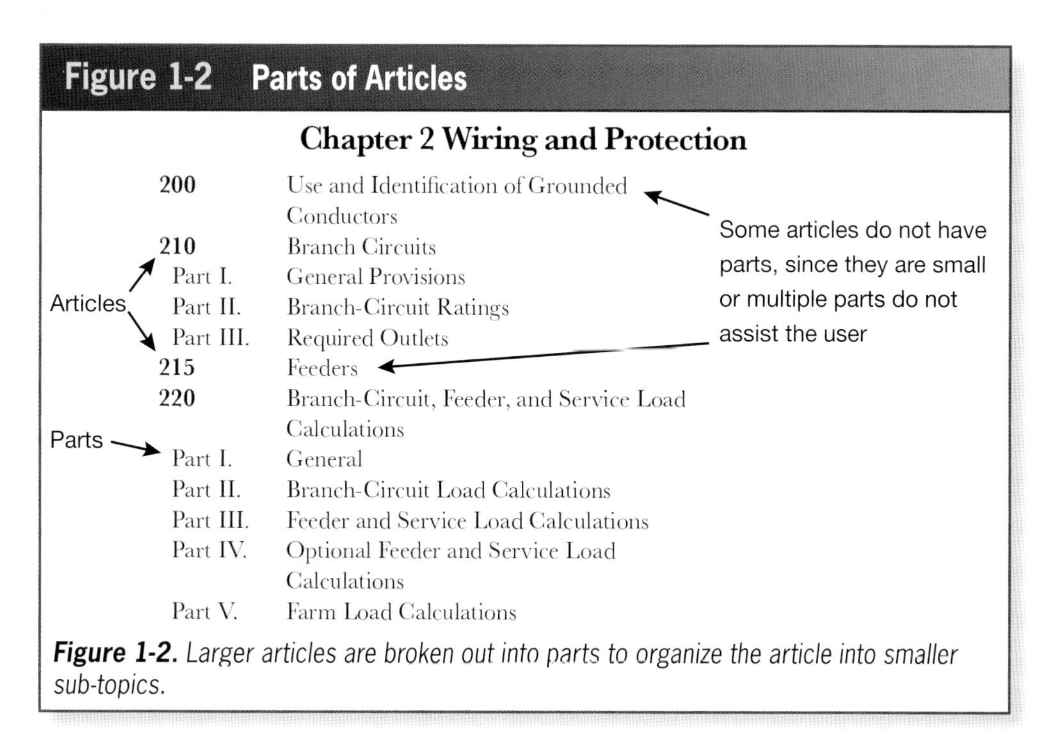

***Figure 1-2.** Larger articles are broken out into parts to organize the article into smaller sub-topics.*

Figure 1-3 Sections and Sub-Sections of Disconnection of an Article

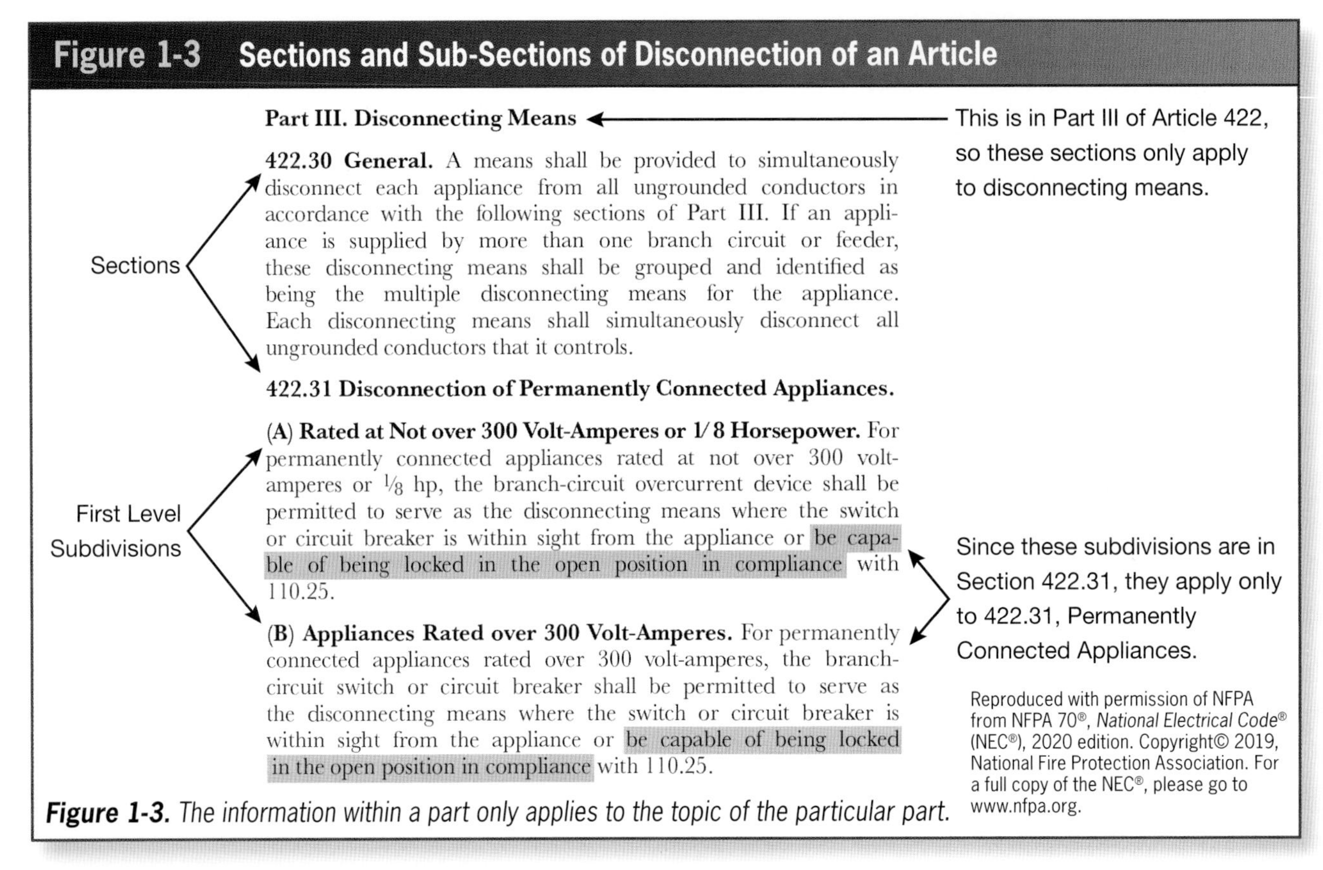

Part III. Disconnecting Means

422.30 General. A means shall be provided to simultaneously disconnect each appliance from all ungrounded conductors in accordance with the following sections of Part III. If an appliance is supplied by more than one branch circuit or feeder, these disconnecting means shall be grouped and identified as being the multiple disconnecting means for the appliance. Each disconnecting means shall simultaneously disconnect all ungrounded conductors that it controls.

422.31 Disconnection of Permanently Connected Appliances.

(A) Rated at Not over 300 Volt-Amperes or 1/8 Horsepower. For permanently connected appliances rated at not over 300 volt-amperes or ⅛ hp, the branch-circuit overcurrent device shall be permitted to serve as the disconnecting means where the switch or circuit breaker is within sight from the appliance or be capable of being locked in the open position in compliance with 110.25.

(B) Appliances Rated over 300 Volt-Amperes. For permanently connected appliances rated over 300 volt-amperes, the branch-circuit switch or circuit breaker shall be permitted to serve as the disconnecting means where the switch or circuit breaker is within sight from the appliance or be capable of being locked in the open position in compliance with 110.25.

Figure 1-3. *The information within a part only applies to the topic of the particular part.*

Figure 1-4 *NEC* Application of Rules

Section	Applies only within the scope of the part of the article in which it is located.
First-Level Subdivision	Applies only within the scope of the section in which it exists.
Second-Level Subdivision	Applies only within the scope of the first-level subdivision in which it exists.
Third-Level Subdivision	Applies only within the scope of the second-level subdivision in which it exists.
List Items	Apply only within the section or subdivision in which they exist.
Exceptions	Apply only to the section, subdivision, or list item under which they exist.
Informational Notes	Used for informational purposes only and are designed to aid the user in the application of the rule(s) under which they exist.

Figure 1-4. *Application of rules in the NEC is in a progressive, ladder-type format.*

Sections

Articles and parts of articles are divided into sections, and each of these sections represents a rule or part of a rule. Sections have bold numbers and a title. Sections may be further broken down into subdivisions. **See Figure 1-3.**

There may be multiple levels of subdivisions or list items. The first-level subdivision is denoted by a bold capital letter, the second-level subdivision by a bold number, and the third-level subdivision by a lowercase letter. It is important to note that in each case they only apply to the section or previous level of subdivision of which they are a part. **See Figure 1-4.**

Subdivisions and Lists

Each section may have up to three levels of subdivisions, and each subdivision may have list items that apply only to the subdivision or section under which they exist. If an item is in bold, it is an article, a first-level subdivision, or a second-level subdivision.

For example, the *NEC* reference for

conductors rated 75°C terminating on equipment rated over 100 amperes is **110.14(C)(1)(b)(1)**, located under **Section 110.14**, first-level subdivision **(C)**, second-level subdivision **(1)**, third-level subdivision **(b)**, list item **(1)**. **See Figure 1-5.**

List items are not limited to being used at a certain level and can be found at any level in the hierarchy.

Exceptions

An exception is used when there is a circumstance or application where a particular requirement does not apply, or where other methods may be used in lieu of the main rule. Exceptions immediately follow the main rule that they modify. If an exception or exceptions apply to items in a numbered list, the exception must indicate which

Figure 1-5 Articles, Sub-Divisions, and Lists

Section	110.14 Electrical Connections
First-Level Subdivision	(A) Terminals
First-Level Subdivision	(B) Splices
First-Level Subdivision	(C) Temperature Limitations
Second-Level Subdivision	(1) Equipment Provisions
Third-Level Subdivision	(a) Temperature Provisions of equipment for circuits rated 100 amperes or less, or marked...
List Items	(1) Conductors rated 60°C (140°F) (2) Conductors with higher temperature ratings...based on 60°C... (3) Conductors with higher temperature ratings if the equipment is listed or identified... (4) For motors marked with design letters....
Third-Level Subdivision	(b) Termination provisions for equipment rated over 100 amperes, or marked...
List Items	(1) Conductors rated 75°C (167°F) (2) Conductors with higher temperature ratings, provided...
Second-Level Subdivision	(2) Separate Connector Provisions
First-Level Subdivision	(D) Installation

Figure 1-5. *Articles are further broken down into subdivisions to further organize the information and guide the user.*

Figure 1-6 Exceptions

358.30 Securing and Supporting. EMT shall be installed as a complete system in accordance with 300.18 and shall be securely fastened in place and supported in accordance with 358.30(A) and (B).

(A) Securely Fastened. EMT shall be securely fastened in place at intervals not to exceed 3 m (10 ft). In addition, each EMT run between termination points shall be securely fastened within 900 mm (3 ft) of each outlet box, junction box, device box, cabinet, conduit body, or other tubing termination.

Exception No. 1: Fastening of unbroken lengths shall be permitted to be increased to a distance of 1.5 m (5 ft) where structural members do not readily permit fastening within 900 mm (3 ft).

Exception No. 2: For concealed work in finished buildings or prefinished wall panels where such securing is impracticable, unbroken lengths (without coupling) of EMT shall be permitted to be fished.

These exceptions apply to 358.30(A). They allow the installer alternate methods for securing EMT within three feet from a junction box.

Figure 1-6. *Exceptions allow for or restrict installation requirements under certain conditions.*

of the list items falls under the exception. Multiple exceptions will be numbered. **See Figure 1-6.**

Informational Notes

Informational notes, according to **90.5(C)**, are explanatory. These provide additional information to clarify the intent or explain the section with which they are associated. These notes are located immediately after the part of the *Code* to which they apply. Informational notes can be text, a table, or an illustration showing an example of an acceptable installation. Informational notes commonly reference outside resources, directing the user to sources where they can find more information regarding a particular application. When a section has more than one informational note, as with exceptions, the notes are numbered. This aids the user in finding and referencing informational notes. Informational notes are not written in mandatory language and do not contain requirements, offer interpretations, or provide recommendations. **See Figure 1-7.**

There are notes below some tables in the *NEC* that are not purely informational. These are mandatory and are to be used with their table.

LANGUAGE OF THE *NEC*

The installer, designer, or engineer who uses the *NEC* must understand how the use of certain words and phrases frames the requirements of the *NEC*. **Section 90.5** explains the differences between mandatory rules and rules that are permissive.

Mandatory Language versus Permissive Language

Mandatory wording is used to explain what must be done to meet a requirement. The word "shall" is used to indicate an action that must be performed. The words "shall not" are used to clearly indicate what actions are prohibited from being performed.

In many instances, users of the *NEC* can clearly see what they shall or shall not do, but they also need to know if certain actions, while not required, are allowed. This is where permissive language is used. Permissive rules are actions that are allowed, but not required. Permissive language is normally used to provide options or different methods available and are indicated by the phrase "shall be permitted." If a particular method or action is not required, the term "shall not be required" is used. These terms are also

Figure 1-7 Informational Notes

Part III. Motor and Branch-Circuit Overload Protection

430.31 General. Part III specifies overload devices intended to protect motors, motor-control apparatus, and motor branch-circuit conductors against excessive heating due to motor overloads and failure to start.

Informational Note No. 1: See Informative Annex D, Example No. D8.

Informational Note No. 2: See the definition of *Overload* in Article 100.

These provisions shall not require overload protection where a power loss would cause a hazard, such as in the case of fire pumps.

Informational Note: For protection of fire pump supply conductors, see 695.7.

Part III shall not apply to motor circuits rated over 1000 volts, nominal.

Informational Note: For over 1000 volts, nominal, see Part XI.

These two informational notes both apply to the first paragraph

These two informational notes both apply to the paragraph immediately above them

Figure 1-7. *Informational notes are not enforceable by the authority having jurisdiction (AHJ). These notes provide additional information that is helpful to the installer.*

used to mandate permission to perform a certain action. If a section of the *Code* were to state that the installer "may" perform an action, an inspector can interpret this as "may or may not." Eliminating this confusion, "Shall be permitted" mandates that permission will be granted, while "Shall not be required" mandates that the action is not required to be done.

The language of the *NEC* is very precise, and vague terms or terms that could be misinterpreted are not included. Such terms include *acceptable, generally, usually,* or *sufficient.*

For example, the rules for establishing the number of receptacles required in a dwelling unit can be found in **210.52**. While it might be technically correct to say that "a reasonable distribution of receptacles prevents the users of the receptacles from resorting to extreme use of extension cords," what is reasonable or extreme to one person may not be seen as reasonable or extreme to another, negating the effectiveness of the requirement. The *NEC* instead uses precise language to indicate the measurements used to create the required receptacle distribution.

Extracted Materials and Other Codes and Standards

There are more than 300 codes and standards published by the National Fire Protection Association. About 8,000 volunteers organized into hundreds of technical committees administer these codes and standards. Each of these standards has the intent to minimize hazards, risks, and the potential for fires. When one NFPA document has jurisdiction over material in the *NEC*, it is extracted and put into the *NEC*. **See Figure 1-8.**

In **Article 517**, for example, there are several extracted items from *NFPA 99: Health Care Facilities Code* and *NFPA 101: Life Safety Code.* This is indicated at the very beginning of the article under the title. After the defintion of "alternate power source" in **517.2**, the reference **[99.3.3.4]** is in brackets, indicating this is extracted from *NFPA 99.* The numeral 3 after 99 indicates

Figure 1-8 NFPA Standards Referenced in the *NEC*

Standard Number	Title	Standard Number	Title
NFPA 20	*Standard for the Installation of Stationary Pumps for Fire Protection*	*NFPA 101*	*Life Safety Code*
NFPA 30	*Flammable and Combustible Liquids Code*	*NFPA 497*	*Recommended Practice for the Classification of Flammable Liquids, Gases, or Vapors and of Hazardous (Classified) Locations for Electrical Installations in Chemical Process Areas*
NFPA 30A	*Code for Motor Fuel Dispensing Facilities and Repair Garages*	*NFPA 499*	*Recommended Practice for the Classification of Combustible Dusts and of Hazardous (Classified) Locations for Electrical Installations in Chemical Process Areas*
NFPA 33	*Standard for Spray Application Using Flammable or Combustible Materials*	*NFPA 501*	*Standard on Manufactured Housing*
NFPA 34	*Standard for Dipping, Coating, and Printing Processes Using Flammable or Combustible Liquids*	*NFPA 790*	*Standard for Competency of Third-Party Field Evaluation Bodies*
NFPA 75	*Standard for the Fire Protection of Information Technology Equipment*	*NFPA 1600*	*Standard on Disaster/Emergency Management and Business Continuity/Continuity of Operations Programs*
NFPA 99	*Health Care Facilities Code*	*NFPA 5000*	*Building Construction and Safety Code*

Figure 1-8. *The* NEC *references other standards, and uses information in other standards to harmonize the* NEC *with those standards.*

that the reference is in **Chapter 3** of *NFPA 99*. In NFPA documents other than the *NEC*, **Chapter 3** is the definitions chapter. These references look different than a reference in the *NEC* because of the style manual used for other NFPA documents. The *NEC* uses the *NEC Style Manual*, while other NFPA documents, such as *NFPA 99*, use the *NFPA Manual of Style*.

Tables

Tables that are not Annex material are enforceable items, as are their associated notes, unless the notes are identified as informational. Many times users new to the *Code* will make the mistake of simply going to a table to find the information they want without remembering one important point: tables generally do not stand alone in the *NEC*. They are attached to the *Code* text that references them. For example, **Table 250.66** is used to size the grounding electrode conductor (GEC). If a person seeking the minimum size GEC simply uses the table without reading the text of **Section 250.66**, he or she may size the GEC incorrectly. The type of grounding electrode used must be considered, as stated in in **250.66**.

Occasionally, a table may be directly referenced, as in **610.42(A)**, where the user is referred to **Table 430.52** when sizing the short-circuit and ground-fault protection of the branch circuit for a crane. This is an important distinction in the *Code* text. In **430.62**, the user is referred to **Section 430.52**, not the table. This allows the use of the table in **430.52** along with all of the attached *Code* language that governs the table. Meanwhile, the reference in **Article 610** only allows the direct use of the table.

430.62 Rating or Setting – Motor Load (A) Specific Load. A feeder supplying a specific fixed motor load(s) and consisting of conductor sizes based on 430.24 shall be provided with a protective device having a rating or setting not greater than the largest rating or setting of the branch-circuit short-circuit and ground-fault protective device for any motor supplied by the feeder [based on the maximum permitted value for the specific type of a protective device in accordance with 430.52, or 440.22(A) for hermetic refrigerant motor-compressors], plus the sum of the full-load currents of the other motors of the group.

610.42 (A) Fuse or Circuit Breaker Rating. Crane, hoist, and mono-rail hoist motor branch circuits shall be protected by fuses or inverse-time circuit breakers that have a rating in accordance with Table 430.52. Where two or more motors operate a single motion, the sum of their nameplate current ratings shall be considered as that of a single motor.

Cross-Reference Tables

The *NEC* has 11 cross-reference tables. **See Figure 1-9.**

There are other tables in **Sections 409.3, 440.3, 680.3, 685.3,** and **705.3.** Notice that (with the exception of **Article 430**) each reference to the section is **XXX.3**. This type of parallel numbering is used where possible in the *NEC* to aid the user in finding particular topics. The topic of **Article 760** is fire alarm systems. **Section 760.3** shows the relation of **Article 760** to other articles in the code. A "dot 3" reference does not always include a cross-reference table. Cross-reference tables point the user to the location where a specific application of a topic is addressed.

Figure 1-9 Cross Reference Tables In The *NEC*

Article 210 Branch Circuits **Table 210.3** Specific-Purpose Branch Circuits
Article 220 Branch-Circuit, Feeder, and Service Load Calculations **Table 220.3** Specific-Purpose Calculation References
Article 225 Outside Branch Circuit and Feeders **Table 225.3** Other Articles
Article 240 Overcurrent Protection **Table 240.3** Other Articles
Article 250 Grounding and Bonding **Table 250.3** Additional Grounding and Bonding Requirements
Article 430 Motors, Motor Circuits, and Controllers **Table 430.5** Other Articles

Figure 1-9. *Several articles in the* NEC *have cross-reference tables to point the user to other requirements that may be relevant to their installation.*

Outlines, Diagrams, and Drawings

The *NEC* does not contain pictures or commentary to aid the *Code* user in the application of installation requirements. As stated in **90.1(C)**, the *NEC* is not intended as a design manual or as an instruction manual for untrained persons. However, the *NEC* does provide some diagrams, drawings, or information in an outline form to guide the user in the correct application of the *NEC* and to further describe the requirements.

The outline drawing in **90.3** is designed to clearly show how the *NEC* is arranged and to show the relationships between the chapters. In **Section 210.52**, the illustration **210.52(C)(1)** shows examples of how the measurement to determine receptacle requirements behind a sink or a range is to be done. **Article 514** has specific areas that are defined as classified (hazardous) locations, and **Figure 514.3** shows where these areas are relative to the fuel dispensers. **See Figure 1-10.** Many other examples of diagrams and figures exist throughout the *NEC*.

Figure 1-10 Illustration of Classified Locations

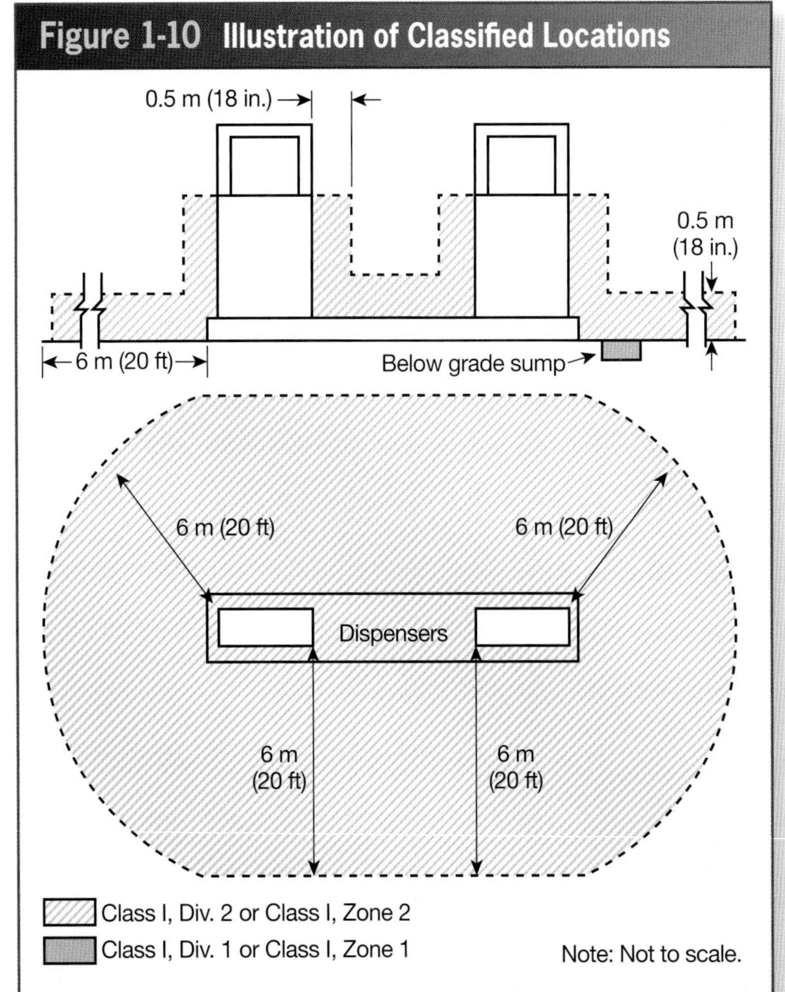

***Figure 1-10.** Requirements for Class I fueling facility environments are covered in **Article 514**.*

The NFPA publishes a handbook for many of their documents. The *NEC Handbook* provides the user with additional information, including drawings and illustrations, along with application notes to aid in further study and understanding of the *NEC*.

CHAPTER 1 OF THE *NEC*

Chapter 1 consists of two articles. **Article 100** provides definitions so *Code* users can apply the requirements within the *NEC* and understand one another when trying to communicate on *Code*-related matters. **Article 110** provides general requirements that Electrical Workers must be familiar with in order to correctly apply the rest of the *NEC*.

Article 100, Definitions

Understanding definitions in the *NEC* is essential to understanding how to use and properly apply the *NEC*. In general, only those terms that are used in two or more articles are defined in **Article 100**. Commonly used terms (terms used in everyday language such as desk, sky, wire, etc.) are not included; however, words such as "Outlet" are included, as they have their own meaning within the *NEC* that differs from the way the word is used commonly. Definitions specific to only one article are included in the article in which they are used and may only be referenced in **Article 100**. Notice that included after each definition is the number of the Code-Making Panel (CMP) that has the primary responsibility for the oversight of that definition.

Article 100 is divided into three parts, each with definitions in alphabetical order.

Part I, General, contains definitions intended to apply wherever the terms are used throughout the *Code*.

Part II, Over 1000 Volts, Nominal, contains definitions applicable only to the parts of articles specifically covering installations and equipment operating at more than a 1000 volts.

Part III, Hazardous (Classified) Locations, contains definitions that apply to hazardous or classified locations. These definitions are used in more than one article in **Chapter 5**, so they are located in **Article 100**.

The *NEC Style Manual* puts forth requirements for defintions in the *NEC* as follows:

- Definitions are in alphbetical order.
- Definitions do not contain the term that is being defined.
- Definitions that appear in two or more articles are listed in **Article 100**.
- Definitions that appear in only one article are in the second section (for example, **690.2**) of that article.

Terms that are used elsewhere might have a significantly different meaning within the context of the *NEC*. All users of the *NEC* must be familiar with the **Article 100** definitions and their meanings in the context of the *NEC*. For example, understanding the differences between branch circuits, feeders, and service conductors is critical, as each circuit type has its own sizing requirements and overcurrent protection requirements.

It is important to remember that, when an Electrical Worker is interacting with customers, the customers will not likely use terms as they are used in the *NEC*. It is up to the Electrical Worker to properly translate the customer's requests into the electrical language of the *NEC*. For example, if a customer asks for twelve outlets on the wall at various locations, he or she is most likely referring to receptacles, but taken literally in *NEC* terms, the customer is actually asking for junction boxes with wiring in them so electrical equipment can be connected at those locations. It is vital to understand the customer's request and to be able to translate his or her meaning into the terms used in the *NEC* to ensure custormer satisfaction and a successful installation.

Many conflicts about *Code* requirements are won or lost over the understanding (or misunderstanding) of definitions. The following definitions are examples of frequently used terms in the *NEC* that the *Code* user must undertand to apply requirements to properly.

Accessible, Readily (Readily Accessible). Capable of being reached quickly for operation, renewal, or inspections without requiring those to whom ready access is requisite to take actions such as to use tools (other than keys), to climb over or under, to remove obstacles or to resort to portable ladders, and so forth. (CMP-1)

The term "readily accessible" is used more than 70 times in the *NEC*. Other terms related to accessibility and defined in **Article 100** are "accessible (as applied to equipment)," "accessible (as applied to wiring methods)," and "concealed."

For example, if a disconnect switch was located behind a stack of boxes, it would be considered *accessible*, but not *readily accessible*. If the disconnect switch had a padlock installed, it would be readily accessible to those personnel who have keys.

Ampacity. The maximum current, in amperes, that a conductor can carry continuously under the conditions of use without exceeding its temperature rating. (CMP-6)

The term "ampacity" is used more than 400 times in the *NEC* and is derived from the combination of the terms "ampere" and "capacity."

Approved. Acceptable to the authority having jurisdiction. (CMP-1)

"Approved" is used more than 300 times in the *NEC*. It is essential that the *Code* user knows that when this term is used, compliance will be determined by the authority having jurisdiction rather than a third party listing organization such as UL, LLC.

The term "authority having jurisdiction" (AHJ) is also defined in **Article 100**. The informational note following the definition states that the AHJ may be a member of federal, state, local, or other regional departments, or an individual such as a fire chief, fire marshal, labor department inspector, health department inspector, electrical inspector, or other individual having statutory authority. The note further states that for insurance purposes, an insurance inspection department, rating bureau, or other insurance company representative may be the AHJ.

Branch Circuit. The circuit conductors between the final overcurrent device protecting the circuit and the outlet(s). (CMP-2)

Branch circuit conductors have different load requirements than feeder or service conductors and are the final pathway serving equipment.

Figure 1-11 Building

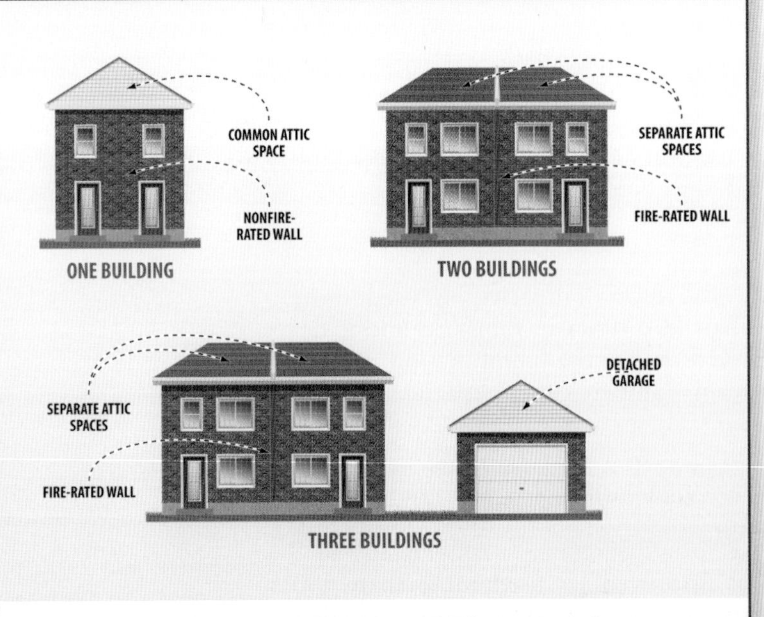

COMMON ATTIC SPACE
NONFIRE-RATED WALL
ONE BUILDING
SEPARATE ATTIC SPACES
FIRE-RATED WALL
TWO BUILDINGS
SEPARATE ATTIC SPACES
FIRE-RATED WALL
DETACHED GARAGE
THREE BUILDINGS

***Figure 1-11.** Buildings are defined in the* NEC *as stand-alone structures or portions of a building separated by fire-rated walls.*

Building. A structure that stands alone or that is cut off from adjoining structures by fire walls with all openings therein protected by approved fire doors. (CMP-1)

The *NEC* contains specific requirements and limitations for buildings, so it is essential to understand that a row of ten strip stores, for example, may be recognized in the *NEC* as 10 separate buildings if they are separated by fire walls. **See Figure 1-11.**

Device. A unit of an electrical system, other than a conductor, that carries or controls electric energy as its principal function. (CMP-1)

The term "device" is used in the *NEC* more than 500 times. Types of devices include, but are not limited to, receptacles and switches. **See Figure 1-12.**

Feeder. All circuit conductors between the service equipment, the source of a separately derived system, or other power supply source and the final branch-circuit overcurrent device. (CMP-2)

For proper application of the *NEC*, the *Code* user must be able to determine the proper *Code* term for all current-carrying conductors. The four types of current-carrying conductors are branch-circuit, feeder, service, and tap conductors.

Fitting. An accessory such as a locknut, bushing, or other part of a wiring system that is intended primarily to perform a mechanical rather than an electrical function. (CMP-1)

Fittings are used to terminate conduit and other raceways to boxes and enclosures.

When the *NEC* requires that equipment such as a motor and an associated

In Sight From (Within Sight From, Within Sight). Where this *Code* specifies that one equipment shall be "in sight from," "within sight from," or "within sight of," and so forth, another equipment, the specified equipment is to be visible and not more than 15 m (50 ft) distant from the other. (CMP-1)

disconnecting means be "within sight of" each other, the requirement is that the equipment be visible and not more than 50 feet apart.

Outlet. A point on the wiring system at which current is taken to supply utilization equipment. (CMP-1)

The term "outlet" would include a receptacle, a ceiling-mounted box for a lighting fixture, and the point at which equipment is hard-wired. **See Figure 1-13.**

Overcurrent. Any current in excess of the rated current of equipment or the ampacity of a conductor. It may result from overload, short circuit, or ground fault. (CMP-10)

An overcurrent includes the following:

- Overload: When the current is above the normal full-load rating, but stays in the normal current path. For example, 23 amperes flowing on a 20-ampere branch circuit is an overload. Too many appliances or equipment may be plugged into the circuit.
- Short circuit: (Not defined in **Article 100** or elsewhere in the *Code*) When the current does not flow through its normal path (takes a shortcut) but continues on the circuit conductors with contact from line-to-line (for example, a black ungrounded conductor touching a red ungrounded conductor of a different circuit) or line-to-neutral contact.

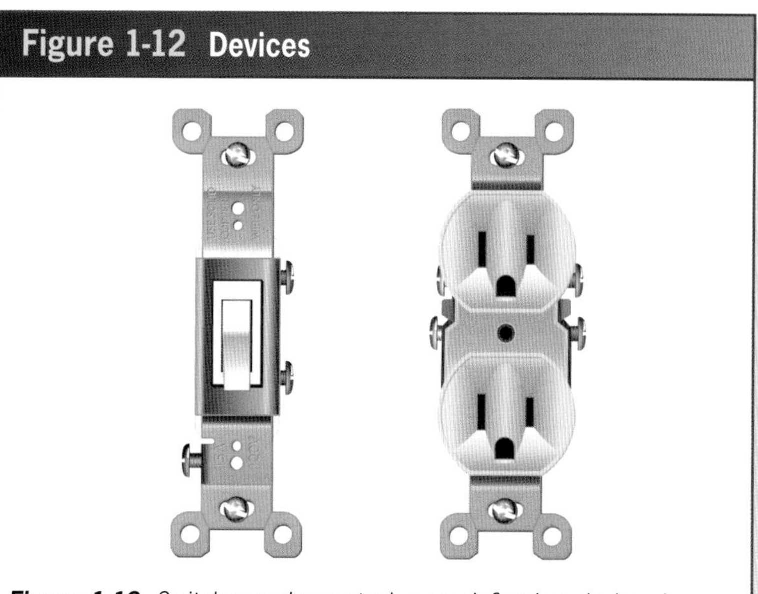

Figure 1-12. *Switches and receptacles are defined as devices in* ***Article 100****.*

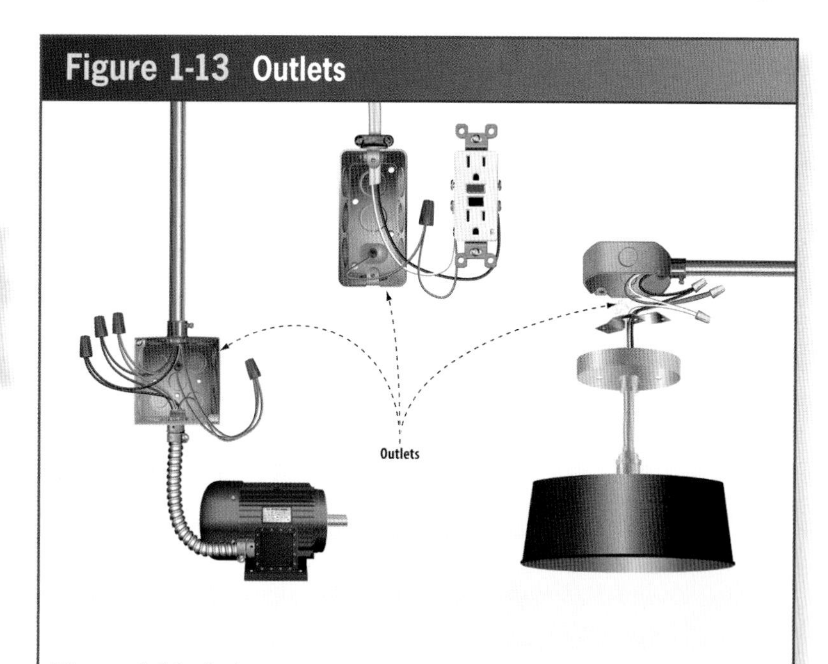

Figure 1-13. *Outlets are not only receptacles, but include lighting outlets in the ceiling.*

Qualified Person. One who has skills and knowledge related to the construction and operation of the electrical equipment and installations and has received safety training to recognize and avoid the hazards involved. (CMP-1)

The term "qualified person" is used more than 100 times in the *NEC*. It is extremely important that all *Code* users understand that a qualified person has specific skills, knowledge, and safety training.

The informational note for this definition refers the user to *NFPA 70E: Standard for Electrical Safety in the Workplace* for electrical safety training requirements.

Article 110, Requirements for Electrical Installations

Article 110 applies to all electrical installations and is divided into five parts. **Part I** is general (applies to all installations), **Part II** applies to installations 1,000 volts or less, and **Part III** applies to those installations over 1,000 volts. The installation and use of portable or mobile high-voltage power distribution and equiment is covered in **Part IV, Tunnel Installations Over 1000 Volts, Nominal**. Some enclosures such as underground vaults have their own unique hazards and challenges, which are covered in **Part V, Manholes and Other Electrical Enclosures Intended for Personnel Entry**.

In **Part I**, the requirements that apply to all installations are presented, regardless of voltages. Every user of the *NEC* must be aware of the requirement in **110.3(B)**, which states that a product must be installed in accordance with the instructions that accompany the product. Listed products are tested with regards to their anticipated use, and installing a product without regard to the instructions may result in an unsafe condition. Electrical connections must be made with care, as when done improperly, connections can fail, resulting in fire, loss of property, and injury to persons. **Section 110.14** holds the requirements to all electrical terminations, including temperature ratings, splices, and torque requirements in **110.14(D)**. In the abscense of manufacturer's recommendations, **Informative Annex I** provides torque values for various screws.

Most equipment that is likely to require examination, adjustment, or servicing while energized must be labeled with arc flash hazard warning labels or stickers that can survive the ambient environment. In addition to the potential risks caused by shock hazards, arc flash can result in severe burns or other physical injuries to the person working on the equipment. **Section 110.16** has the requirements to warn qualified persons of the potential arc flash hazards. **Sections 110.21**, **110.22**, and **110.24** contain marking and identification requirements for equipment, disconnects, and service equipment.

Perhaps one of the most frequently-used tables in the *NEC* is found in **Part II** of **Article 110**. **Table 110.26(A)(1)** in **Section 110.26** contains the requirements for clear working distances around equipment. Other important requirements for height, illumination, and width are also located in **110.26**. Making a mistake and having to perform a correction to comply with working distance requirements is very costly, so it is imperative that users of the *NEC* consult this section when planning an installation.

The last three parts of **Article 110** are more specialized and contain requirements for high-voltage, tunnel, and manhole installations.

HIGHLIGHTING AND PREPARING THE *CODE* BOOK

Finding information quickly in the *NEC* is important to success as a student and as a worker. Marking and annotating the *Code* book can be very helpful in finding information. The following suggestions will make any copy of the *NEC* easier to use and will enhance the reader's ability to quickly and accurately find the information they need.

Highlighting can be very useful for *Code* users. In fact, certain portions of the *Code* book have already been highlighted by the publisher. Gray highlighting is used for each article title. In the text body of a section, gray highlighting is also used to indicate any changes from a previous edition of the *NEC*. The most important rule for highlighting is to highlight only where absolutely necessary. It is counterproductive to highlight an entire page of the *Code* book.

Fact

The colors chosen to be used to highlight a *Code* book are up to the individual user of the *NEC*. Once colors have been chosen, stay consistent with how the color is used.

Before Marking a *Code* Book...

If the *Code* book is going to be used on an exam for licensing or certification, verify what highlighting, notes, tabs, or mark-ups, if any, are allowed. Some jurisdictions may even require using a *Code* book they supply. Requirements vary from jurisdiction to jurisdiction.

Open the *NEC* to the table of contents. The table of contents is extremely useful in finding information in the *NEC*. Highlighting it will make it easier to use. Using a green highlighter, highlight each article. Notice that most articles in the table of contents have a high voltage section. Highlight these parts in the table of contents with a different color highlighter, such as orange. If a question ever arises regarding higher voltages, these entries will be easier to find in the Table of Contents.

Chapter 3 of the *NEC* covers topics such as cable assemblies, raceways, and support methods. Each of the wiring methods typically have their own acronym (NMC, IGS, MI, etc.). If the user is not thoroughly familiar with each acronym, highlighting the acronyms in the table of contents with a yellow highlighter may be helpful.

Continue highlighting the table of contents with each article in green. After highlighting the table of contents, move on to the rest of the *Code* book. Highlight the title of each article in green if desired, then find the parts of each article and highlight them in orange. Being able to quickly find the appropriate part of an article is imperative.

Another suggested practice is to use yellow highlighting or a red pen to underline text, subdivisions, key words, list items, and any additional information that may be useful in the future. This method helps find previously studied material and is extremely useful when revisiting specific areas of the *Code*.

Chapter 1 General

Using different colors for highlighting is recommended

Highlighting acronyms makes them easier to find and match to their reference.

ARTICLE 110
Requirements for Electrical Installations

Part I. General

110.1 Scope. This article covers general requirements for the examination and approval, installation and use, access to and spaces about electrical conductors and equipment; enclosures intended for personnel entry; and tunnel installations.

Consistently using the same color for highlighting titles and parts is important.

Exception: Individual conductors shall be permitted where installed as separate overhead conductors in accordance with 225.6.

(B) Conductors of the Same Circuit. All conductors of the same circuit and, where used, the grounded conductor and all equipment grounding conductors and bonding conductors shall be contained within the same raceway, auxiliary gutter, cable tray, cablebus assembly, trench, cable, or cord, unless otherwise permitted in accordance with 300.3(B)(1) through (B)(4).

(1) Paralleled Installations. Conductors shall be permitted to be run in parallel in accordance with the provisions of 310.10(G). The requirement to run all circuit conductors within the same raceway, auxiliary gutter, cable tray, trench,

Good highlighting produces a reference that is customized to the user's needs.

SUMMARY

The *Code* is arranged so **Chapters 1** through **4** apply generally to all installations and **Chapters 5** through **7** address "special" installations with requirements that can be more restrictive than found in **Chapters 1** through **4**. **Chapter 8**, Communications, does not apply to **Chapters 1** to **7** unless it specifically references an article or section in **Chapters 1** to **7**, and **Chapter 9** provides tables referenced in other articles. This arrangement is detailed in **Section 90.3**.

The *NEC* is divided into 10 major parts: the introduction and nine chapters. The *Code* is organized into chapters, parts, articles, sections, subsections, exceptions, and informational notes common to installation requirements. Parts are major subdivisions of articles that logically separate information for ease of use and proper application.

Parts are then broken down into separate sections and individually titled to address the scope of the individual part. Sections may be logically subdivided into as many as three levels. Sections may also contain list items, exceptions, and informational notes. Understanding the language and definitions used in the *Code* is vital to properly interpreting the sections.

The structure of the *NEC* is in a progressive, ladder-type format, which when applied is as follows:

- The rule in the **second-level subdivision**, which applies only under:
  - The rule in the **first-level subdivision**, which applies only under:
    - The **section** in which it exists, which applies only in:
      - The **part** it is arranged in, which applies only in:
        - The **article** in which it exists, which is limited to:
          - The **chapter** in which it is located.
- **List items** are used in sections, subdivisions, or exceptions where necessary.
- **Exceptions** are used only when absolutely necessary and are always italicized.
- **Informational notes** are informational only and are not mandatory.
- **Tables** in **Chapter 9** are applicable only where referenced elsewhere in the *NEC*.
- **Informative Annexes** are informational only and are not mandatory.

An in-depth understanding of the structure or outline form of the *NEC* is necessary to properly apply the requirements of the *Code*.

In accordance with **90.3**, **Chapter 1** applies to all electrical installations. **Article 100** holds defintions that are found in two or more articles in the *NEC*. Other definitions specific to the articles they are in are found in those articles. Understanding definitions is a key element to understanding the *NEC*.

Article 110, Requirements for Electrical Installations, has information relevant to all installations. More specific requirements that pertain to specific equipment or environments are found in their respective articles.

If done in an organized manner, highlighting the *NEC* enhances the information and contents to further assist the user to quickly find the information they need.

QUIZ

Answer the questions below using a copy of the *NEC* and by reviewing Chapter 1 of this text.

1. **If a definition is located in an article other than Article 100, what section is it placed in?**
   a. XXX.2
   b. XXX.3
   c. XXX.4
   d. XXX.5

2. **Section __?__ explains how the code is organized and what chapters apply generally and what chapters are special and modify the other chapters of the *NEC*.**
   a. **90.1**
   b. **90.3**
   c. **100**
   d. **110.1**

3. **What part of Article 110 is specific to installations above 1,000 volts?**
   a. **Part I**
   b. **Part II**
   c. **Part III**
   d. **Part IV**

4. **A point on the wiring system in which current is taken to supply utilization equipment is known as a(n) __?__.**
   a. fitting
   b. junction box
   c. outlet
   d. receptacle

5. **Conductors used to normally carry current in the *NEC* are copper or aluminum. If the material is not specified, the sizes given are for __?__ conductors.**
   a. copper
   b. aluminum
   c. copper or aluminum
   d. equipment grounding

6. **Which chapter of the *NEC* is considered to stand by itself?**
   a. **Chapter 1**
   b. **Chapter 8**
   c. **Chapter 9**
   d. Informative Annexes

7. **The purpose of the *NEC* is to safeguard persons and property. Compliance with the *NEC* will __?__.**
   a. eliminate all hazards arising from the use of electricity
   b. ensure an installation that is convenient
   c. provide for future expansion
   d. provide an installation that is essentially free from hazards

8. **Which of the following is not covered by the *NEC*?**
   a. Recreational vehicles
   b. Mobile homes
   c. Installations in mines
   d. Both a. and c.

9. **Which Annex provides torque values for electrical connections in the absence of manufacturer's recommendations?**
   a. **Annex A**
   b. **Annex C**
   c. **Annex I**
   d. **Annex T**

10. **One inch is equal to 25.4 millimeters. This is an example of a __?__ conversion.**
    a. fair
    b. firm
    c. hard
    d. soft

Planning the Installation

Chapter 2 of the *NEC* contains information that is essential to all electrical installations. Before conductors, raceways, and equipment can be installed, the installation must be planned. **Chapter 2** covers branch circuits, feeders, and services, which are common to all occupancies. Also covered are important safety elements such as overcurrent protection for conductors and equipment, grounding and bonding, and overvoltage protection.

Chapter 2 and **Chapter 3** of the *NEC* form the backbone of the *Code*. These two chapters cover the planning and the installation of electrical distribution and power systems. Before the installation, however, the electrical loads must be calculated, the size of conductors must be determined, and the grounding system must be planned. This information is all covered in **Chapter 2** of the *NEC*.

Objectives

- Recall the titles of the articles found in **Chapter 2** of the *NEC*.
- Locate definitions applicable to **Chapter 2** of the *NEC*.
- Apply articles, parts, and tables to identify specific *NEC* **Chapter 2** requirements.

Chapter

Table of Contents

CHAPTER 2 OF THE *NEC*

Chapter 2 is a core section of the *NEC* and applies to all installations, as stated in **Section 90.3**. **Chapter 2** explains the requirements for the major circuit types found in the *NEC*. There are nine articles in **Chapter 2**. Seven of these articles relate to types of circuits, and two of these articles cover general rules about conductor protection from overcurrents, overvoltages, and surges.

These circuit types are defined in **Article 100** and **Article 240**, and are discussed extensively in **Chapter 2** of the *NEC*.

> **Grounded Conductor.** A system or circuit conductor that is intentionally grounded.

In many cases, but not all, the grounded conductor is also the neutral conductor of a system.

> **Branch Circuit.** The circuit conductors between the final overcurrent device protecting the circuit and the outlet(s).

Branch circuits are the most common circuits installed. When equipment or a load is fed by a panel, a branch circuit will likely be used.

> **Feeder.** All circuit conductors between the service equipment, the source of a separately derived system, or other power supply source and the final branch-circuit overcurrent device.

Feeders can feed a sub-panel, for example. **See Figure 2-1.**

Figure 2-1 Circuit Types in Electrical Systems

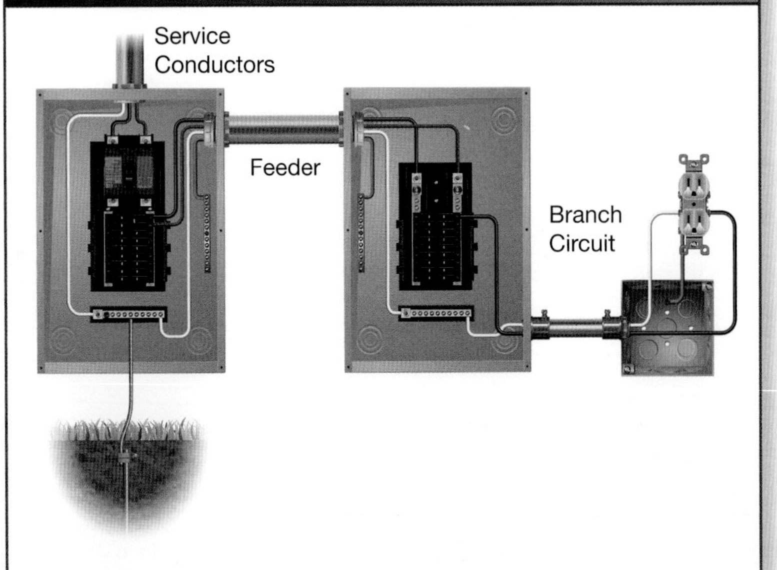

Figure 2-1. *Each circuit type has its own article in the* NEC.

> **Service.** The conductors and equipment connecting the serving utility to the wiring system of the premises served.

> **Tap Conductors.** Defined in Article 240 as a conductor, other than a service conductor, that has overcurrent protection ahead of its point of supply that exceeds the value permitted for similar conductors that are protected as described elsewhere in 240.4.

There are other specialized circuits discussed in the *NEC* pertaining to special applications or conditions such as sensitive electronic equipment, signaling, and lower voltage uses. These are covered in the "Specials" section (**Chapters 5**, **6**, and 7) of the *NEC*.

Article 200, Use and Identification of Grounded Conductors

Note the scope of **Article 200**. This article provides requirements for grounded conductors, their identification, and the identification of their terminals.

The grounded conductor is intentionally grounded at the service or at a separately derived system, depending on the type of system installed, and provides a reference to ground (earth) for the electrical system. The grounded conductor can be a phase conductor or a neutral conductor. Systems that would use a neutral conductor would include the nominal voltages of 120/208-volt, 120/240-volt, 277/480-volt, and 347/600-volt systems. Systems with only one voltage, such as 240 volts or 480 volts, are either ungrounded or will have one phase conductor grounded. These grounded conductors are not neutral conductors by definition.

When speaking with individuals not familiar with electrical systems, the term *grounded conductor* is frequently misused and often mistakenly equated with a *grounding conductor*. The grounded conductor can carry fault current like a grounding conductor, but the grounded conductor is intended to carry current used by the load during normal operation, while the grounding conductor is not. This is another reason familiarity with the *NEC* is essential. Frequently, the Electrical Worker has to translate the language a customer uses into *Code* language to ensure a proper installation while still meeting the customer's requirements and desires.

When the grounded conductor is used as a neutral conductor, **Article 200** requires it to be grouped with the ungrounded or "hot" conductors with which the neutral is associated. This prevents the neutral from being used for more than one circuit from each phase, thereby avoiding overloading the neutral conductor.

Article 200 also contains requirements for how to identify a grounded conductor. Generally, the *NEC* does not have requirements for identifying or color-coding conductors with specific colors. However, **Article 200** does call for specific identification of the grounded conductor; for this reason, the grounded conductor is frequently referred to as the *identified conductor*.

Grounded conductors are to be white or gray in color. When more than one voltage system is used, the grounded conductor must be uniquely identified; an example would be using gray for one system and white for the other. See **200.6(D)**. Further identification may be achieved by striping the conductor with three white or gray stripes, but this is not required unless more than two system voltages are present in the premises wiring systems.

When identifying conductors 6 AWG and smaller, the entire conductor must be identified, but when conductors are 4 AWG or larger, the identification is allowed to be applied only at termination points and where the conductors are accessible. From a *Code* standpoint, it is important to note that this wire size requirement for identification applies *only* to grounded conductors, as stated in **Article 200**. This is why it is important to pay attention to the scope of each article.

Terminals for grounded conductors are required to be identified according to **Section 200.10**. The method of identification varies by device and application.

Article 210, Branch Circuits

Article 210 provides general requirements for branch circuits. The key word here is *general*. For example, other branch circuits for specific loads, such as motors and appliances, have their own requirements according to their respective articles. These other types of branch circuits are addressed in **Table 210.3**. **See Figure 2-2.**

Fact

A short circuit occurs when current flows between phases, or from an ungrounded "hot" conductor to the grounded or equipment grounding conductor, where the load is bypassed.

Figure 2-2 Table 210.3

Table 210.3 Specific-Purpose Branch Circuits

Equipment	Article	Section
Air-conditioning and refrigerating equipment		440.6, 440.31, 440.32
Busways		368.17
Central heating equipment other than fixed electric space-heating equipment		422.12
Fixed electric heating equipment for pipelines and vessels		427.4
Fixed electric space-heating equipment		424.3
Fixed outdoor electrical deicing and snow-melting equipment		426.4
Infrared lamp industrial heating equipment		422.48, 424.3
Motors, motor circuits, and controllers	430	
Switchboards and panelboards		408.52

***Figure 2-2.** Tables such **Table 210.3** in the Code guide the user to other relevant references.*

Fact

While many colors may be chosen for ungrounded conductors, **Sections 200.6** and **250.119** are mandatory requirements for grounded and equipment grounding conductors.

Section 210.5 lists the methods and requirements for identification of conductors. Proper and consistent identification is necessary for safe installations. The *NEC* provides rules for identification and the posting of the identification method in **210.5**. This section does not mention specific colors, and while certain color combinations may be common in practice—such as using black, red, and blue for ungrounded phase conductors—this is not required. The rules for identifying direct current (DC) systems and circuits are found in **210.5(C)(2)**.

The use of branch circuits is limited to certain voltages by the load being operated and the occupancy. For example, 277 volts is a common voltage for commercial lighting, but it is not permitted in dwelling units. See *NEC* 210.6(A).

The majority of ground-fault circuit interrupter (GFCI) requirements are found in **Section 210.8**. The *NEC* has addressed GFCI receptacles since the 1968 edition, but the locations where GFCI protection is required have since been expanded or modified. **Section 210.8** is a good example of why it is important for the user to understand how the *Code* is organized.

GFCI receptacles in dwelling units are covered in **210.8(A)**, while **210.8(B)** is concerned with locations other than dwelling units. **See Figure 2-3.** Notice that **210.8(C)**, which covers lighting outlets in crawl spaces, would apply to all crawl spaces regardless of the type of occupancy in which they are found. Since these requirements are in **Chapter 2**, they apply generally to all sections. Other articles, such as **Article 547, Agricultural Buildings**, and **Article 680, Swimming Pools, Fountains, and Similar Installations** have additional requirements.

The *NEC* introduced arc-fault circuit interrupter (AFCI) protection in **Section 210.12** in 2002. As with GFCI protection, AFCI protection has since been expanded beyond its original application, which was bedroom outlets. Other sections in **210.12** cover dormitories, guest rooms, and suites, and include requirements for modifying existing branch circuits.

Figure 2-3 Countertop Receptacle

Figure 2-3. *Receptacles that serve kitchen countertops in a dwelling unit must be GFCI protected per* ***210.8(A)(6)****.*

*The small appliance circuits required in **210.11(C)(1)** also serve receptacles in dining rooms.*

Part II of **Article 210** discusses the ratings of branch circuits and the required ratings of receptacles and devices used with them. The rating of a branch circuit is the rating of the overcurrent protective device, such as a circuit breaker, protecting the conductors. **Table 210.24** offers a summary of branch circuit requirements, but for complete information, consult the full text of the *Code*.

Sections 210.19 and **210.20** cover the minimum ampacity rating of conductors and overcurrent protection of branch circuits. Conductors must carry the load and be protected by the fuse or circuit breaker that supplies them.

The third part of **Article 210** covers receptacle locations for dwelling units and other required electrical outlets. This information may seem out of place to new users of the *NEC* until the title of **Part III, Required Outlets**, is considered. An outlet is a point on the electrical system that supplies current to utilization equipment, receptacles, and so forth. Planning the branch circuit requires the designer to know where the circuit must be routed and what it is required to supply.

Section 210.52 contains the rules for planning receptacle locations in dwelling units. Other required outlets for servicing equipment, storage areas, and meeting rooms conclude **Article 210**.

Foyers have special spacing requirements for receptacles.

Article 215, Feeders

After the designer has determined the necessary branch circuits and the requirements to supply them, he or she must then determine the feeder or service size. **Article 215** is a shorter article, but it offers prescriptive guidance on the proper sizing, protection, and identification of feeder conductors. Much of this language is similar to the language used for branch circuits, as feeders use many of the same components, albeit on a larger scale.

Article 220, Branch-Circuit, Feeder, and Service Load Calculations

Article 220 covers how to calculate the loads for branch circuit, feeder, and service applications.

Proper application of the requirements depends on paying careful attention to the parts of this article.

Part I	**General**
Part II	**Branch-Circuit Load Calculations**
Part III	**Feeder and Service Load Calculations**
Part IV	**Optional Feeder and Service Load Calculations**
Part V	**Farm Load Calculations**

Other articles in the first four chapters of the *NEC* that apply to calculations are referenced in **Table 220.3**. However, other references to consider in **Chapters 5**, **6**, and **7** are not listed in this table. Per **Section 90.3**, the requirements in these chapters would still apply and thereby modify **Chapter 2**.

Lighting loads are calculated based upon square footage and occupancy type. The 2020 edition of the *NEC* incorporates information from energy code requirements. New technology and increasing concerns over energy use, combined with local codes requiring the designer to comply with energy use regulations, led to the modification of **Table 220.12**.

Section 220.14 provides a checklist of sorts to guide the installer in calculating the branch circuit, feeder, and service loads.

In **Part III** of **Article 220**, feeder and service load calculations are discussed. This part of the article includes multiple tables and rules for applying demand factors. For example, **Table 220.54** addresses demand factors for household clothes dryers. Notice that as the number of dryers increases, the demand percentage decreases. If an apartment complex has 10 units, each with its own dryer, then the demand is only 50% of the total. This is because it is understood that in normal use, all ten tenants of the apartment complex are unlikely to run their dryers at the same time.

The application of a demand factor is a common modification made to services and feeders due to the combination of loads present and the load diversity in a panel. **Article 220** discusses situations when the application of a demand factor is permitted for lighting, receptacles, kitchen equipment, ranges, and so forth.

Parts II and **III** of **Article 220** cover what is called "the standard method" for calculating loads. **Part IV** covers the optional method. The optional method is also allowed, and frequently—but not always—results in a lower value for load requirements. The neutral conductor for either method is still calculated the same way, by consulting the appropriate table in **Article 250** for the minimum size in all cases and then consulting **Section 220.61** to size the neutral for the load present. **Part IV** also includes requirements for schools and restaurants.

Part V of **Article 220** exclusively covers the calculation of loads for farms. Farms are unique in that they typically have both a dwelling unit and other loads found in commercial or industrial applications.

Article 225, Outside Branch Circuits and Feeders

Article 225 contains the requirements unique to installing branch circuits and feeders outside of a building. Services are not included in **Article 225**, since services are covered separately in **Article 230** and service conductors are generally required to be outside of a building. It is critical not to overlook requirements in **Article 225** when an installation moves outside.

Article 230, Services

In order to understand services, it is necessary to understand several of the definitions in **Article 100**.

Service. The conductors and equipment connecting the serving utility to the wiring system of the premises served.

Service Conductors. The conductors from the service point to the service disconnecting means.

Service Conductors, Overhead. The overhead conductors between the service point and the first point of connection to the service-entrance conductors at the building or other structure.

Service Conductors, Underground. The underground conductors between the service point and the first point of connection to the service-entrance conductors in a terminal box, meter, or other enclosure, inside or outside the building wall.

Service Drop. The overhead conductors between the serving utility and the service point.

Service-Entrance Conductors, Overhead System. The service conductors between the terminals of the service equipment and a point usually outside the building, clear of building walls, where joined by tap or splice to the service drop or overhead service conductors.

Service-Entrance Conductors, Underground System. The service conductors between the terminals of the service equipment and the point of connection to the service lateral or underground service conductors.

Service Equipment. The necessary equipment, consisting of a circuit breaker(s) or switch(es) and fuse(s) and their accessories, connected to the serving utility, and intended to constitute the main control and disconnect of the serving utility.

Service Lateral. The underground conductors between the utility electric supply system and the service point.

Service Point. The point of connection between the facilities of the serving utility and the premises wiring.

Figure 230.1 in the *NEC* is useful in finding the appropriate section under **Services.**

Service conductors are different from feeders or branch circuits in that they are unfused and have no protection. This is why they are installed mostly outside of buildings and only allowed inside buildings to a limited extent. The disconnect to a building's service must be located either outside the building or inside the building near the point where the service conductors enter the building. See **Section 230.70** for more information.

The primary purpose of the *Code* is to protect people and property from the hazards that can result from the use of electricity. Therefore, conductors that are not protected by an overcurrent protective device are usually installed outside the building, as with a fuse, or limited in length, as with tap conductors.

Following **Part I** of **Article 230**, which is general and applies to all services, **Parts II** and **III** contain the requirements for overhead and underground service conductors, respectively. The size, rating, and methods of wiring are all covered.

Service entrance conductors are addressed in **Part IV** of **Article 230**. This part covers the allowed methods, sizes, and rating requirements.

Fact

Overcurrent protective devices must also be rated for the fault current that is available on the circuit they are protecting. See **Section 110.9**.

Part V covers general service equipment requirements, which include guarding, marking, and surge protection.

Part VI discusses the disconnecting means, including the maximum number of disconnecting means required in an installation and where they must be located. The service of a building must be located and arranged such that the power to the building can be shut off quickly and easily. The requirements in **Sections 230.70** to **230.85** must be followed. **Section 230.85** in the 2020 edition of the *NEC* includes a new requirement. For one- and two-family dwelling units, an emergency disconnect must be installed that is readily accessible outside of the dwelling unit. **See 230.85** for a discussion of the three allowed options.

When service conductors enter a building, they are terminated on an overcurrent protective device (OCPD) such as a circuit breaker or fuse. These requirements are covered in **Part VII, Service Equipment — Overcurrent Protection**. **Section 230.90** contains the requirements for sizing the OCPD. Note **Exception No. 2**, which in some cases allows the OCPD to have a rating higher than the ampacity of the service conductors. This is usually not the case with other unfused conductors such as tap conductors.

Article 240, Overcurrent Protection

Article 240 contains the requirements for protecting conductors against overcurrent. Other articles, such as those covering services, branch circuits, motors, and fire pumps, may have their own protection requirements that reference or augment **Article 240**. It is important to consult all the articles in the *Code* that are relevant to the installation.

Part I of **Article 240** contains the requirements for the OCPDs for cords and conductors. Standard OCPD sizes can be found in **Section 240.6**. **Table 240.4(G)** points to other references used to size overcurrent protection.

It is important not to confuse the overcurrent protection of a conductor with the ampacity of a conductor. While the two topics are related to each other, they are very different. A conductor may be required to be protected by a 15-ampere overcurrent device, but may actually have a higher ampacity; such is the case with a 14 AWG conductor on a 15-ampere circuit breaker. This leads many new users of the *NEC* to incorrectly assume that a 14 AWG conductor is only suitable for 15 amperes. In fact, **Article 240** is limiting the use of the conductor. As always, check other references to ensure **Article 240** is being applied correctly.

Part I requires the OCPD to be sized to protect the ungrounded conductors, while **Part II** generally requires the OCPD to be at the point of supply of the ungrounded conductors in **Section 240.21**. The rules starting with **240.21(B)** are commonly called the "tap rules." These rules allow a conductor to be installed where not protected by an upstream fuse or circuit breaker under very limited circumstances of size, length, and method.

The locations where fuses or circuit breakers must be installed on the premises are also covered in **Part II**. Overcurrent protective devices must generally be readily accessible and not any higher than six feet and seven inches from the floor. Additionally, since fault currents are involved with the operation of overcurrent protective devices, they must not be located where combustible items are stored.

Article 242, Overvoltage Protection

In the 2020 edition of the *NEC*, **Articles 280** and **285** have been combined into one article. Now overvoltage and surge protection both below and above 1,000 volts are addressed in the same article. This more closely resembles the rest of the *Code*, which is organized first by topic and then into more specific sections.

High-voltage surge arresters protect equipment from lightning or electrical faults, and are commonly installed on the load side of the service equipment. Lower-voltage surge suppressors are used on grounded circuits 1,000 volts or less.

Article 250, Grounding and Bonding

Article 250 is the longest article in the *Code*. An understanding of how it is organized, when combined with the Codeology method, will help the user find what he or she is looking for within the article. **Figure 250.1** may also be helpful to the reader in sorting out the many parts of **Article 250**. **See Figure 2-4.**

As is the case with other articles in the *Code*, **Article 250** begins with the scope and general requirements of the article. The purpose and key elements of **Article 250** are explained in **250.4(A)(1)** through **(5)**. **Part II** continues with an overview of system-level requirements, such as the systems and voltages required to be grounded and the sizing of the conductors that accomplish the connection to ground. More specific system types, such as DC systems and systems over 1,000 volts, are discussed further in **Parts VIII** and **X**.

Once the requirements for grounding have been established in **Part II**, **Part III** addresses how the connection to ground should be made. The grounding electrode is the connecting element to earth. There can be, and usually is, more than one electrode; when connected together they form the grounding electrode system. This system of electrodes is connected together and to the incoming electrical system of the building, whether it is a service or a feeder, via bonding jumpers or grounding electrode conductors.

Part III is organized in the sequence of planning the installation of a grounding system. **Section 250.52** contains the requirements for the electrodes themselves, including the physical dimensions, configurations, and materials of the electrodes. **Section 250.53** moves on to the installation requirements for the various electrodes, including the installation methods used and other considerations such as burial depth and spacing from other electrodes. The pattern of specifying the dimensions, materials, and installation is repeated with the grounding electrode conductor. **Section 250.62** covers the construction and materials of the conductor, while **250.64** introduces the installation requirements of the grounding electrode conductor. **Part III** continues with the proper

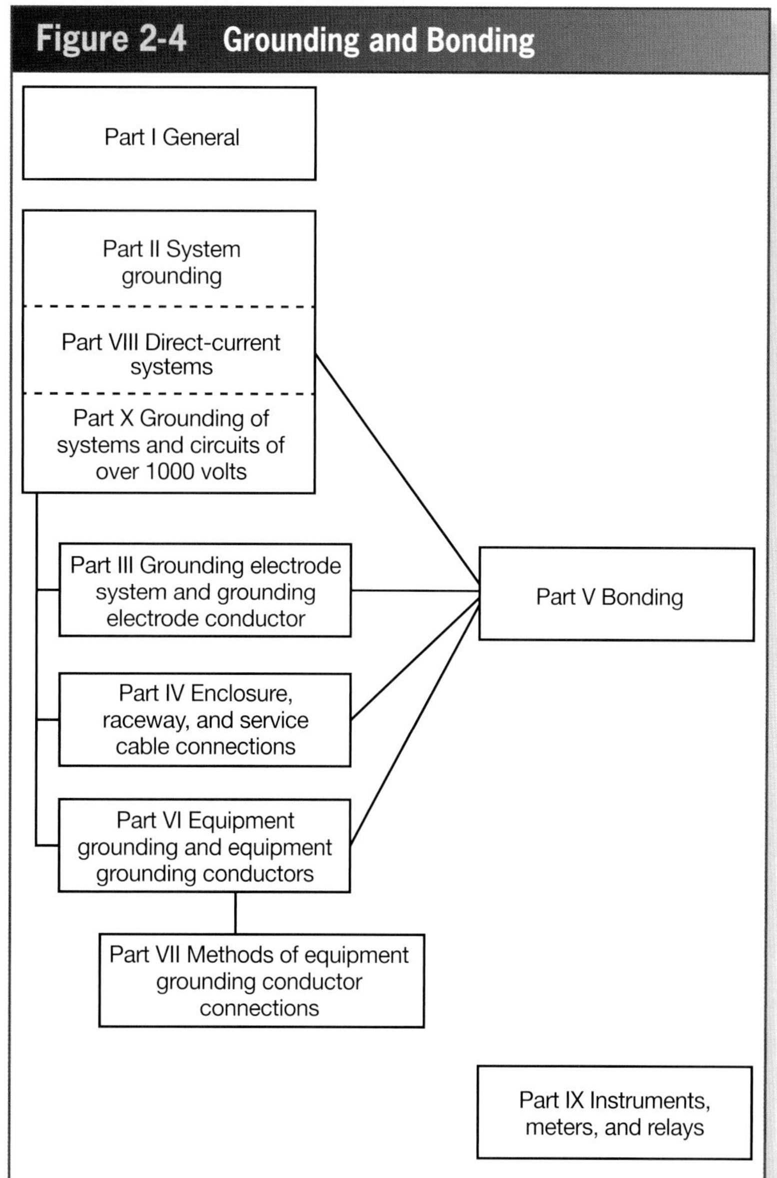

***Figure 2-4.** A figure from **Section 250.1** aids Code users in finding the relevant Parts in **Article 250**.*

sizing of the grounding electrode conductor in **Section 250.66**.

Part IV of **Article 250** is a short but important part containing the requirements for connecting enclosures and raceways. This discussion leads into the closely related topic of bonding, covered in **Part V**. This part of **Article 250** establishes the requirement that all metal enclosures and raceways must be connected to the grounded service conductor or, in the case of an ungrounded system, connected to the grounding electrode conductor. This connection ensures that all enclosures in an electrical system have the same potential as each other and that they all have the same potential as the ground.

When items such as raceways, enclosures, and equipment are electrically continuous, they are said to be bonded together. There is a general requirement in the *NEC* for these items to be bonded. If they cannot be connected directly, an equipment bonding jumper and conductor are required.

Section 250.92 has special requirements for the bonding of services. Fault currents at services can be very high, and it is important for a bonding means to not only be an effective return path to operate overcurrent protective devices, but for it to survive to do the job again, if needed.

The 2008 edition of the *NEC* saw the introduction of the intersystem

Once the concrete is poured for a foundation, the previously-installed rebar will serve as a concrete-encased electrode per ***250.52(A)(3)****. Note the green sticker indicating approval of the AHJ.*

bonding termination device, discussed in **250.94**. This device offers an easily accessible way to connect the grounding means of a communication system to the building's electrical system grounding connection. Prior to this, communication systems were often bonded to water pipes, which may or may not have been connected to the grounding system due to replacement of metal piping in the building. The intersystem device provides a consistent and reliable location for these connections.

Table 250.102(C) is used to size the grounded or bonding conductors at the service or on the secondary side of a transformer. When there is a fault on a service upstream of the main circuit breaker or fuse, the only limiting factor under the control of the *NEC* is the size of the conductors. The size of the grounded conductor is dependent upon the size of the ungrounded or "hot" conductors, as shown in the table.

Part VI states what equipment is required to be grounded and defines what can serve as the equipment grounding conductor. **Section 250.118** is a very important section in **Part VI**. It defines what can be used as an equipment grounding conductor, what raceways can be used, whether any raceways have restrictions, and any additional requirements for the equipment grounding conductor. Whenever a conductor is used as an equipment grounding conductor, the installer should reference **Section 250.122.**

Table 250.122 in **Part VI** is also used for sizing conductors, similar to **250.102(C)**, with one key difference. A fault on these conductors is limited by an OCPD such as a circuit breaker or fuse. Be sure to refer to the sub-sections of **250.122**. An equipment grounding conductor is never required to be larger than the ungrounded conductors supplying the equipment.

As was required in **Part IV**, most items shall be bonded or made electrically continuous. **Part VII** of **Article 250** contains the requirements for the grounding of equipment. This can be accomplished by connecting the equipment to supports that are grounded, to an enclosure that is grounded via a raceway such as those mentioned in **250.118**, or to the equipment grounding conductor.

In the past, grounding of DC systems has not been a major topic of concern. The increase in the use and integration of photovoltaic systems, microgrids, and wind power has changed this. **Part VIII** of **Article 250** contains special requirements for the grounding of DC systems.

The last two parts of **Article 250** address the grounding of instruments and relays along with the grounding of systems over 1,000 volts. Note that whenever a system over 1,000 volts is being installed, **Article 490** must also be consulted.

SUMMARY

Section 90.3 states that **Chapter 2** of the *NEC* applies generally to all installations. Proper planning of any installation is required in order for safe and efficient systems to be put into use. The *NEC* has specific identification requirements for the grounded conductors in **Article 200** and the grounding conductors in **Article 250**. The grounded conductor, also known as the identified conductor, is commonly identified with white or gray markings or insulation. **Article 210, Branch Circuits**, has requirements specific to branch circuits. Commonly referenced topics and sections include:

- GFCI Protection – **Section 210.8**
- AFCI Protection – **Section 210.12**
- Branch Circuit Minimum Ampacity – **Section 210.19**
- Overcurrent Protection of Branch Circuits – **Section 210.20**
- Outlet Requirements for General Use, Specific Areas and Equipment – **Sections 210.52**, **210.63**, and **210.70**

After the branch circuit needs are determined, **Article 215, Feeders**, or **Article 230, Services**, should be brought into the planning process. These articles have specific requirements for sizing and protecting their conductors. Services, covered in **Article 230**, are typically outside or underground. **Article 225** contains the additional requirements for branch circuits and feeders when installed outside of buildings.

Overvoltage protection is the scope of **Article 242**. Overvoltage or surge protection can be installed at points from the branch circuit all the way to the service conductors of a building.

Grounding and bonding of systems and equipment is essential to the safety of any electrical installation. **Article 250** covers grounding and bonding, and contains many references with which an Electrical Worker must be familiar. **Figure 250.1** in the *NEC* provides a convenient reference to the parts of **Article 250**. Being familiar with the parts of **Article 250** will go a long way in helping users of the *NEC* find the information they are looking for.

Chapter 2 is one of the four cornerstones upon which the *NEC* is built. **Chapters 1** through **4** form the foundation of the *NEC*, and any well-built foundation requires proper planning.

QUIZ

Answer the questions below using a copy of the *NEC* and by reviewing Chapter 2 of this text.

1. **Which of the following would be an acceptable means of identifying the grounded conductor of an AC system?**
   a. Black insulation with three white stripes along its entire length
   b. Gray insulation
   c. White insulation
   d. Any of the above
2. **If the ungrounded service conductors in an installation are 3/0 AWG, what is the minimum size of the grounded conductor?**
   a. 1/0 AWG
   b. 1 AWG
   c. 2 AWG
   d. 4 AWG
3. **Fill in the correct article number or the title of the article.**
   __?__ - Article 210
   Grounding and Bonding - Article __?__
   Services - Article __?__
   __?__ - Article 215
4. **Which article covers protection from overvoltages and surges?**
   a. **Article 200**
   b. **Article 220**
   c. **Article 242**
   d. **Article 280**
5. **What is the reference used to size the grounding electrode for a DC system?**
   a. **250.20**
   b. **250.66**
   c. **250.102**
   d. **250.166**
6. **The overcurrent protection for a branch circuit must be at least __?__.**
   a. 80% of the noncontinuous load
   b. 100% of the noncontinuous load
   c. 125% of the continuous load
   d. both b. and c.
7. **In a dwelling, receptacles in which of the following locations must be GFCI protected?**
   a. Basements
   b. Kitchen countertops
   c. Outdoor receptacles
   d. All of the above
8. **The __?__ is/are the underground conductors from the utility up to the point of connection to the premises wiring.**
   a. service conductors
   b. service lateral
   c. service point
   d. underground service conductors
9. **A Type 1 surge protective device (SPD) is not permitted to be installed __?__.**
   a. in circuits over 1,000 volts
   b. on the supply side of the service disconnect
   c. anywhere on the load side of the service disconnect overcurrent device
   d. none of the above
10. **Table __?__ is used to size the smallest grounded conductor of a feeder allowed.**
    a. **250.66**
    b. **250.102(C)**
    c. **250.122**
    d. none of the above

Building the Installation

Chapter 3, Wiring Methods and Materials, is perhaps the chapter an Electrical Worker references most frequently when installing parts of an electrical system. **Chapter 3** is used when building and installing an electrical system. It includes everything from general conductor and raceway requirements in the beginning of the chapter to more detailed information regarding ampacity, support, box fill, and conduit fill. While some of these components are chosen in the planning stages, they are all implemented in the building phase of a project. Each of the circuits will, in turn, be connected to equipment, motors, switches, receptacles, lighting fixtures, and so forth. Equipment requirements are found in **Chapter 4** of the *NEC*.

Objectives

» Recall the titles of the articles found within **Chapter 3** of the *NEC*.

» Locate definitions applicable to **Chapter 3** of the *NEC*.

» Identify the general requirements for wiring methods and materials.

» Recognize conductors for general wiring.

» Apply articles, parts, and tables to identify specific *NEC* **Chapter 3** requirements.

Chapter

Table of Contents

CHAPTER 3 OF THE *NEC*

The 47 articles in **Chapter 3** of the *National Electrical Code* form one of the four foundational corners on which the *NEC* is built. **Chapter 3, Wiring Methods and Materials**, applies to all installations unless modified by **Chapters 5, 6**, or **7**. These chapters follow the pattern exhibited elsewhere in the *NEC*, starting with general applications and moving on to topics that are more specific. In these chapters, the topics are related to wiring, sizing and construction of conductors, raceway and cable assemblies, enclosures, support methods for conductors, and more. All of these items have one thing in common: they are critical to installing and building an electrical installation.

Article 300, Wiring Methods

General requirements for the use of wiring methods are found in **Article 300**. There are two parts to **Article 300**; **Part I** is general and **Part II** contains requirements for installations over 1,000 volts. Requirements specific to certain raceway types are found in the article covering the type of raceway in question. For example, **300.11** requires raceways to be securely fastened in place, which applies to all raceways. However, if the raceway being installed is electrical metallic tubing, specific support and securing requirements are found in **Article 358**.

Part I of **Article 300** holds sections of *Code* that all Electrical Workers must be familiar with. **Section 300.2** indicates that the methods used in **Chapter 3** shall be used for installations 1,000 volts or less unless permitted elsewhere in the chapter, such as in **Article 311, Medium Voltage Conductors and Cable**. Conductors over 1,000 volts are permitted elsewhere in the *NEC*, as well.

Section 300.3 specifies when conductors may be run as single conductors or parallel conductors, and under what conditions conductors from different systems may share the same raceway or enclosure. The general rule that conductors of the same circuit must be installed in the same raceway, cable assembly, or trench is addressed in **300.3(B)**.

Sections 300.4 and **300.5** reveal how to protect conductors, raceways, and cables installed aboveground and underground, respectively. When installing cable such as nonmetallic sheathed cable in bored holes through wood studs, care must be taken to bore the hole in the proper place or protect the cable with a steel plate, commonly called a nail plate. This is covered in **300.4**. Another aboveground concern involves installing boxes, cables, or raceways under metal-corrugated sheet roof decking. Per **300.4(E)**, the electrical installation must be installed one and a half inches from the roof decking to prevent damage from screws that may be used to repair or replace the decking. Notice the exception allows rigid and intermediate metal conduit. **See Figure 3-1.**

Figure 3-1 Requirements in 300.4(E)

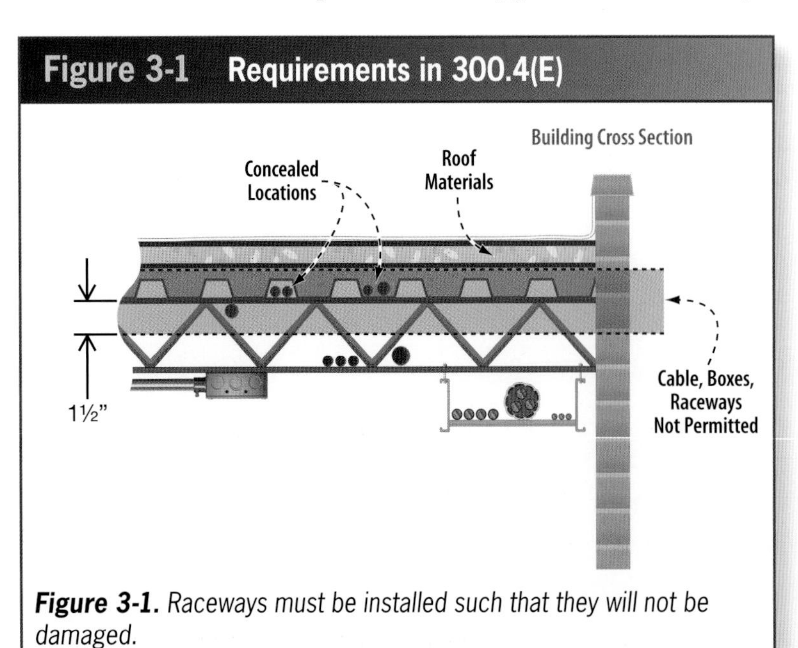

Figure 3-1. *Raceways must be installed such that they will not be damaged.*

Requirements for underground installations are covered in **300.5**. **Table 300.5** shows the proper burial depths for raceways, cable assemblies, and circuit types for various environments or areas. **See Figure 3-2.** Remember that the notes to a table are just as much a part of the table as the contents of the table itself.

Figure 3-2 Table 300.5 Minimum Cover Requirements

Table 300.5 Minimum Cover Requirements, 0 to 1000 Volts, Nominal, Burial in Millimeters (Inches)

Location of Wiring Method or Circuit	Type of Wiring Method or Circuit									
	Column 1 Direct Burial Cables or Conductors		Column 2 Rigid Metal Conduit or Intermediate Metal Conduit		Column 3 Nonmetallic Raceways Listed for Direct Burial Without Concrete Encasement or Other Approved Raceways		Column 4 Residential Branch Circuits Rated 120 Volts or Less with GFCI Protection and Maximum Overcurrent Protection of 20 Amperes		Column 5 Circuits for Control of Irrigation and Landscape Lighting Limited to Not More Than 30 Volts and Installed with Type UF or in Other Identified Cable or Raceway	
	mm	in.	mm	in.	mm	in.	mm	in.	mm	in.
All locations not specified below	600	24	150	6	450	18	300	12	150[a, b]	6[a, b]
In trench below 50 mm (2 in.) thick concrete or equivalent	450	18	150	6	300	12	150	6	150	6
Under a building	0 (in raceway or Type MC or Type MI cable identified for direct burial)	0	0	0	0	0	0 (in raceway or Type MC or Type MI cable identified for direct burial)	0	0 (in raceway or Type MC or Type MI cable identified for direct burial)	0
Under minimum of 102 mm (4 in.) thick concrete exterior slab with no vehicular traffic and the slab extending not less than 152 mm (6 in.) beyond the underground installation	450	18	100	4	100	4	150 (direct burial) 100 (in raceway)	6 4	150 (direct burial) 100 (in raceway)	6 4
Under streets, highways, roads, alleys, driveways, and parking lots	600	24	600	24	600	24	600	24	600	24
One- and two-family dwelling driveways and outdoor parking areas, and used only for dwelling-related purposes	450	18	450	18	450	18	300	12	450	18
In or under airport runways, including adjacent areas where trespassing prohibited	450	18	450	18	450	18	450	18	450	18

[a]A lesser depth shall be permitted where specified in the installation instructions of a listed low-voltage lighting system.

[b]A depth of 150 mm (6 in.) shall be permitted for pool, spa, and fountain lighting, installed in a nonmetallic raceway, limited to not more than 30 volts where part of a listed low-voltage lighting system.

Notes:

1. Cover is defined as the shortest distance in mm (in.) measured between a point on the top surface of any direct-buried conductor, cable, conduit, or other raceway and the top surface of finished grade, concrete, or similar cover.
2. Raceways approved for burial only where concrete encased shall require concrete envelope not less than 50 mm (2 in.) thick.
3. Lesser depths shall be permitted where cables and conductors rise for terminations or splices or where access is otherwise required.
4. Where one of the wiring method types listed in Columns 1 through 3 is used for one of the circuit types in Columns 4 and 5, the shallowest depth of burial shall be permitted.
5. Where solid rock prevents compliance with the cover depths specified in this table, the wiring shall be installed in a metal raceway, or a nonmetallic raceway permitted for direct burial. The raceways shall be covered by a minimum of 50 mm (2 in.) of concrete extending down to rock.

Figure 3-2. Table 300.5 *shows the minimum amounts of cover required above a wiring method installed underground, according to the area of installation.*

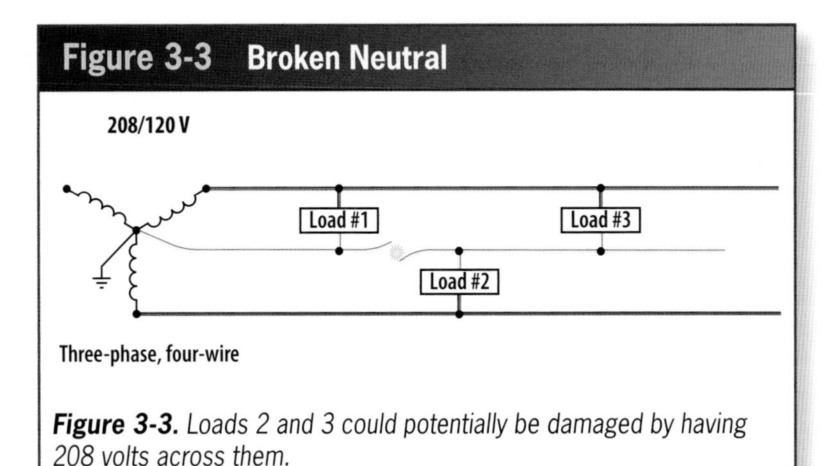

Figure 3-3. *Loads 2 and 3 could potentially be damaged by having 208 volts across them.*

> **Fact**
>
> Safety is a very important concern when installing conductors vertically. If the conductors are being fed from a higher height to a lower height, the weight of the conductors may become too much to control and may fall down the raceway, pulling the conductors off the reels. On the other hand, if the conductors are being pulled from above, the rope may break, resulting in a similarly difficult situation.

Other environmental considerations include corrosive atmospheres, variances in temperature, and installation of conductors with non-electrical systems. These concerns are addressed in **Sections 300.6** through **300.9**.

Section 300.11 covers the requirements for securing and supporting raceways and cables. Raceways cannot generally be used to support cables, and the *NEC* makes it clear that cables cannot be used to support raceways, either. See **300.11(C)** and **(D)**. When installing a raceway, the raceway must be electrically and mechanically continuous, and the same is true for conductors. Per **300.13(B)**, a device is not allowed to provide continuity to the grounded conductor of a multiwire circuit, since the removal of such a device could cause a break in the neutral and result in a hazardous situation leading to damaged equipment. **See Figure 3-3.**

When splices or connections to devices are made, they must be made in a box or conduit body, unless the splice or connection is made according to one of the items in **310.15(A)** through **(L)**. The number of conductors in an enclosure is covered separately in **Article 314**. Installing conductors in a vertical raceway invites additional concerns, as addressed in **Section 300.19**. The weight of the conductors adds stress to the terminations, and must be managed using the support methods explained in **300.19**.

Part II of **Article 300** provides additional rules for high-voltage conductors more than 1,000 volts. The majority of **Part II** covers underground installations. See **300.50** for these requirements.

Article 310, Conductors for General Wiring

Article 310 covers requirements for conductors including ampacity, insulation ratings, conductor types, marking, and identification. These requirements do not apply to conductors that are a part of equipment, but rather to conductors that are installed as part of the premises wiring system. Other special conductor requirements are covered elsewhere in the *NEC*, such as flexible cords and cables in **Article 400** and fixture wires in **Article 402**. Unless derating is required, **Article 310** generally does not apply to the applications of flexible cords, cables, or fixture wires.

Beginning with the 2020 edition of the *NEC*, the scope of **Article 300** includes conductors up to and including 2,000 volts. Higher voltages that were previously covered in **Article 310** have been moved to the new **Article 311**. This differentiation helps the user quickly find the requirements for a particular installation, making the *NEC* easier to use.

There are three parts in **Article 310**:

- **Part I, General**
- **Part II, Construction Specifications**
- **Part III, Installation**

Part I covers the requirements for all conductors 2,000 volts and less. Conductors may be solid or stranded, although conductors size 8 AWG and larger installed in raceways are generally required to be stranded. Conductors may be made of aluminum or copper, which have a minimum size of 12 AWG or 14 AWG, respectively.

Part II starts with **Section 310.4** and the associated table, **Table 310.4(A)**. **See Figure 3-4.** This table contains the specifications for insulated conductors, while **310.4(B)** covers the insulation specifications for conductors rated 2,000 volts. Everyone involved in an electrical installation, from the installer

Figure 3-4 Conductor Constructions and Applications

Table 310.4(A) Conductor Applications and Insulations Rated 600 Volts[1]

Trade Name	Type Letter	Maximum Operating Temperature	Application Provisions	Insulation	Thickness of Insulation: AWG or kcmil	Thickness of Insulation: mm	Thickness of Insulation: mils	Outer Covering[2]
Fluorinated ethylene propylene	FEP or FEPB	90°C (194°F)	Dry and damp locations	Fluorinated ethylene propylene	14–10 8–2	0.51 0.76	20 30	None
		200°C (392°F)	Dry locations — special applications[3]	Fluorinated ethylene propylene	14–8	0.36	14	Glass braid
					6–2	0.36	14	Glass or other suitable braid material
Mineral insulation (metal sheathed)	MI	90°C (194°F) 250°C (482°F)	Dry and wet locations For special applications[3]	Magnesium oxide	18–16[4] 16–10 9–4 3–500	0.58 0.91 1.27 1.40	23 36 50 55	Copper or alloy steel
Moisture-, heat-, and oil-resistant thermoplastic	MTW	60°C (140°F) 90°C (194°F)	Machine tool wiring in wet locations Machine tool wiring in dry locations. Informational Note: See NFPA 79.	Flame-retardant, moisture-, heat-, and oil-resistant thermoplastic	 22–12 10 8 6 4–2 1–4/0 213–500 501–1000	(A) (B) 0.76 0.38 0.76 0.51 1.14 0.76 1.52 0.76 1.52 1.02 2.03 1.27 2.41 1.52 2.79 1.78	(A) (B) 30 15 30 20 45 30 60 30 60 40 80 50 95 60 110 70	(A) None (B) Nylon jacket or equivalent
Paper		85°C (185°F)	For underground service conductors, or by special permission	Paper				Lead sheath
Perfluoro-alkoxy	PFA	90°C (194°F) 200°C (392°F)	Dry and damp locations Dry locations — special applications[3]	Perfluoro-alkoxy	14–10 8–2 1–4/0	0.51 0.76 1.14	20 30 45	None
Perfluoro-alkoxy	PFAH	250°C (482°F)	Dry locations only. Only for leads within apparatus or within raceways connected to apparatus (nickel or nickel-coated copper only)	Perfluoro-alkoxy	14–10 8–2 1–4/0	0.51 0.76 1.14	20 30 45	None
Thermoset	RHH	90°C (194°F)	Dry and damp locations		14-10 8–2 1–4/0 213–500 501–1000 1001–2000	1.14 1.52 2.03 2.41 2.79 3.18	45 60 80 95 110 125	Moisture-resistant, flame-retardant, nonmetallic covering[2]
Moisture-resistant thermoset	RHW RHW-2	75°C (167°F) 90°C (194°F)	Dry and wet locations	Flame-retardant, moisture-resistant thermoset	14–10 8–2 1–4/0 213–500 501–1000 1001–2000	1.14 1.52 2.03 2.41 2.79 3.18	45 60 80 95 110 125	Moisture-resistant, flame-retardant, nonmetallic covering
Silicone	SA	90°C (194°F) 200°C (392°F)	Dry and damp locations For special application[3]	Silicone rubber	14–10 8–2 1–4/0 213–500 501–1000 1001–2000	1.14 1.52 2.03 2.41 2.79 3.18	45 60 80 95 110 125	Glass or other suitable braid material

***Figure 3-4. Table 310.4** includes technical information about the types of conductors that may be used for general wiring applications. See the NEC to review this table in its entirety.*

Figure 3-5 Letter Designations for Insulation Characteristics

Letter Designation	Meaning
B	Braid
FEP	Fluorinated ethylene propylene insulation
H	75°C (Lack of "H" indicates 60°C)
N	Nylon jacket
PFA	Perfluoroalkoxy insulation
R	Thermoset insulation
S	Silicone insulation
T	Thermoplastic (if the first letter)
TFE	Polytetrafluoroethylene
U	Underground use
W	Moisture resistant
X	Cross-linked polymer insulation
Z	Modified ethylene tetrafluoroethylene insulation

Figure 3-5. *Electrical Workers must know the limitations and requirements for different insulation types.*

to the designer, must be aware of the limitations of the different conductor types found in **Table 310.4(A)**. **See Figure 3-5.** Notice that according to the table, some types of insulation are allowed to be used in wet environments. This is important even if the conductor is in a raceway, as addressed in **Section 300.9**.

Section 310.6 requires that ungrounded conductors, commonly called "hot" conductors, be identified in a manner different than that for grounded or grounding conductors. While it is common practice to identify the lower system voltages in a 120/208-volt AC system with black, red, and blue for the three different phases, this is not a requirement in the *NEC*. Similarly, there is no requirement for brown, orange, and yellow to be used on 277/480-volt systems. The only references to a specific color code for multiple phases are located in the special chapters (**Chapters 5**, **6**, and **7**) of the *Code*. For example, the requirement for identifying the phase conductors of an isolated power system can be found in **517.160(A)(5)**.

Another case where color coding is used to identify a phase conductor is

Reminder

The special chapters modify **Chapters 1** through **7** per **Section 90.3**.

Fact

Isolated power systems are used in hospitals (a type of special occupancy) in critical procedure locations such as operating rooms to limit any ground-fault current to a low value while not interrupting the power supply. Interruptions of electrical power to an operating room cannot be tolerated.

covered in Sections **110.15** and **230.56**. In a 4-wire delta system, such as a 120/240-volt system, only two of the three phases will have a nominal voltage to the grounded conductor. One of the three phases will have a higher voltage of about 208 volts to ground, or neutral. **Sections 110.15** and **230.56** require this conductor, called a "high leg," to be identified using the color orange. This color is used to alert the electrician not to connect any 120-volt equipment to this phase, as such a connection would result in the destruction of the equipment.

Part III of **Article 310** is one of the most frequently used portions of the *NEC*. **Section 310.10** covers the uses permitted for conductors in various environments and applications. Per **310.10(G)**, conductors in parallel are generally required to have a minimum size of 1/0 AWG. Parallel conductors must also meet the following requirements:

- They must be terminated in the same manner.
- They must consist of the same conductor material.
- They must be the same size in circular mil area.
- They must have the same insulation type.
- They must be the same length.

The requirement that parallel conductors must be the same length prevents one conductor from becoming overloaded due to the nature of current in a parallel circuit. The *NEC* does not specify how much of a difference in length is acceptable, if any. This allowance would be determined by the authority having jurisdiction.

The service and feeder conductors for a dwelling unit are sized differently than other service or feeder conductors. While other service and feeder conductors are sized based upon the size of the load and the overcurrent protective device using the ampacity tables referenced in **310.15**, these applications for dwelling units are sized based on the service or feeder rating alone. See **310.12**.

The ampacity of the conductors is sized based upon 83% of this service or feeder rating. Notice that this 83% factor can only be used when the service or feeder supplies the entire dwelling and associated loads. If a 100-ampere feeder is used to supply power to an accessory building, then the 83% factor may not be used. **Table 310.12** shows the sizes permitted to be used with the 83% factor included. **See Figure 3-6.**

The ampacity of a conductor is the maximum amount of current a conductor can carry continuously under the conditions of use without exceeding its temperature rating. **Sections 310.14** and **310.15** contain requirements that must be considered when determining the size of a conductor in terms of the load being served.

When the ampacity is not calculated under engineering supervision as covered in **310.14**, **Section 310.15** can be used along with **Tables 310.16** through

Reminder

The *NEC* is not an instruction manual for untrained persons. To become a skilled Electrical Worker, knowledge of all system types is necessary.

Figure 3-6 Table 310.12

Table 310.12 Single-Phase Dwelling Services and Feeders

Service or Feeder Rating (Amperes)	Conductor (AWG or kcmil)	
	Copper	Aluminum or Copper-Clad Aluminum
100	4	2
110	3	1
125	2	1/0
150	1	2/0
175	1/0	3/0
200	2/0	4/0
225	3/0	250
250	4/0	300
300	250	350
350	350	500
400	400	600

Note: If no adjustment or correction factors are required, this table shall be permitted to be applied.

Figure 3-6. Table 310.12 *is used to size service or feeder conductors where they serve the entire dwelling unit and associated loads.*

Figure 3-7 Medium Voltage Cable

Figure 3-7. The shielding around medium-voltage (MV) cable is used as a ground reference and to distribute voltage stresses around the conductor in a uniform manner.

Courtesy of Okonite Cable Company

310.20. Conductor ampacity is modified by such variables as:

- Number of current-carrying conductors in raceway, or other installation considerations
- Ambient temperature
- Temperature rating of the terminations
- Conductor material
- Type of insulation

Determining ampacity is a complex process that must be fully understood by every Electrical Worker.

Fact

What is commonly called a panel is actually an enclosure that has a panelboard within it. The panelboard is the assembly comprised of bus bars and that allow for the connection of the circuit breakers. When looking for an item in the *Code* book, it is important the user understands the question in the terms used by the *Code*. This is why understanding definitions is so important.

Article 311, Medium Voltage Conductors and Cable

In the 2020 edition of the *NEC*, the information from Article 328, Medium Voltage Cable, was moved to the new **Article 311**.

Article 311 covers the installation, construction, and use of medium-voltage conductors and cables. Medium-voltage cable is referred to as Type MV. Type MV cable can be a single- or multi-conductor cable with a solid dielectric (insulator) rated from 2,001 volts to 35,000 volts. For the most part, these voltages are seen as the purview of utilities, but larger facilities may have services with higher voltages. These facilities may include hospitals, water treatment plants, manufacturing facilities, and data centers, among others. **Article 311** specifies that the same items that influence lower voltage conductors in **Article 310** can have similar effects on MV conductors.

Part III of **Article 311** discusses the installation requirements of medium-voltage conductors and cables. For several editions, the *NEC* has called for qualified persons to be the ones who install, terminate, and test MV cable. In the 2020 *NEC*, this requirement, along with other information from Article 328, Medium Voltage Cable, was moved to the new **Article 311**.

When installing MV cable, the installer must consider the pulling tension on the conductors and the sidewall pressure on the conductors as they are pulled through bends in the raceway. Proper lubricants compatible with the jacket of the cable must be used, and the bending radii—both the dynamic radius when the conductor is being installed, and the static radius once the installation is complete—must be considered. Medium-voltage cable has unique components such as shielding that must not be damaged during or after installation. **See Figure 3-7.**

Article 312, Cabinets, Cutout Boxes, and Meter Socket Enclosures

The scope of **Article 312** covers the installation and construction of cabinets, cutout boxes, and meter socket enclosures. This includes providing adequate room for conductors within these enclosures. According to **Section 312.6**, conductors must not be deflected (bent) without having adequate space. As a result, this article has influence on sizing junction boxes, wireways, and gutters. **Article 312** is referenced in **Articles 300**, **314**, **366**, **376**, and **378**.

Bending large conductors in a small space can put undue stress on terminations and insulation. Providing space for wiring and splices in a panel is required, as shown in **312.8**. Take note of the language in **312.8, Switch and Overcurrent Device Enclosures**.

Article 314, Outlet, Device, Pull, and Junction Boxes; Conduit Bodies; Fittings; and Handhole Enclosures

The first three parts of **Article 314, Scope and General**, **Installation,** and **Construction Specifications**, respectively, hold the requirements for supporting, sizing, and installing junction boxes, enclosures, and conduit bodies installed in systems of 1,000 volts or less. **Part II** of **Article 314** is the part of this article most frequently referenced by the Electrical Worker or designer of the electrical system.

Section 314.16 in **Part II** is used to properly size the boxes needed for conductors, devices, and fittings. Each conductor that enters a junction box is counted according to **314.16(B)(1)**. **Table 314.16(A)** includes the standard size boxes and the cubic inches provided. **See Figure 3-8.** Manufacturers may stamp a different value on their boxes, but if a manufacturer's stamp is not present, the table value must be used.

The volume that a device requires is calculated based upon how much space the device takes up and the size of the conductors required to be connected to the device, according to **314.16(B)(4)**. Proper sizing of junction boxes will result in a smoother installation that will minimize heat, reduce ground faults, and prevent damage to the conductors.

Figure 3-8 Table 314.16(A) Metal Boxes

Table 314.16(A) Metal Boxes

Box Trade Size			Minimum Volume		Maximum Number of Conductors* (arranged by AWG size)						
mm	in.		cm³	in.³	18	16	14	12	10	8	6
100 × 32	(4 × 1¼)	round/octagonal	205	12.5	8	7	6	5	5	5	2
100 × 38	(4 × 1½)	round/octagonal	254	15.5	10	8	7	6	6	5	3
100 × 54	(4 × 2⅛)	round/octagonal	353	21.5	14	12	10	9	8	7	4
100 × 32	(4× 1¼)	square	295	18.0	12	10	9	8	7	6	3
100 × 38	(4 × 1½)	square	344	21.0	14	12	10	9	8	7	4
100 × 54	(4 × 2⅛)	square	497	30.3	20	17	15	13	12	10	6
120 × 32	(4¹¹⁄₁₆ × 1¼)	square	418	25.5	17	14	12	11	10	8	5
120 × 38	(4¹¹⁄₁₆ × 1½)	square	484	29.5	19	16	14	13	11	9	5
120 × 54	(4¹¹⁄₁₆ × 2⅛)	square	689	42.0	28	24	21	18	16	14	8
75 × 50 × 38	(3 × 2 × 1½)	device	123	7.5	5	4	3	3	3	2	1
75 × 50 × 50	(3 × 2 × 2)	device	164	10.0	6	5	5	4	4	3	2
75× 50 × 57	(3× 2 × 2¼)	device	172	10.5	7	6	5	4	4	3	2
75 × 50 × 65	(3 × 2 × 2½)	device	205	12.5	8	7	6	5	5	4	2
75 × 50 × 70	(3 × 2 × 2¾)	device	230	14.0	9	8	7	6	5	4	2
75 × 50 × 90	(3 × 2 × 3½)	device	295	18.0	12	10	9	8	7	6	3
100 × 54 × 38	(4 × 2⅛ × 1½)	device	169	10.3	6	5	5	4	4	3	2
100 × 54 × 48	(4 × 2⅛ × 1⅞)	device	213	13.0	8	7	6	5	5	4	2
100 × 54 × 54	(4 × 2⅛ × 2⅛)	device	238	14.5	9	8	7	6	5	4	2
95 × 50 × 65	(3¾ × 2 × 2½)	masonry box/gang	230	14.0	9	8	7	6	5	4	2
95 × 50 × 90	(3¾ × 2 × 3½)	masonry box/gang	344	21.0	14	12	10	9	8	7	4
min. 44.5 depth	FS — single cover/gang (1¾)		221	13.5	9	7	6	6	5	4	2
min. 60.3 depth	FD — single cover/gang (2⅜)		295	18.0	12	10	9	8	7	6	3
min. 44.5 depth	FS — multiple cover/gang (1¾)		295	18.0	12	10	9	8	7	6	3
min. 60.3 depth	FD — multiple cover/gang (2⅜)		395	24.0	16	13	12	10	9	8	4

*Where no volume allowances are required by 314.16(B)(2) through (B)(5).

Figure 3-8. Table 314.16(A) *includes the minimum volume allowance for metal boxes, along with the number of conductors allowed where they are all the same size.*

Section 314.16 contains the requirements for enclosures that contain conductors 6 AWG and smaller. For enclosures that contain conductors larger than 6 AWG, **Section 314.28** is used.

CABLES, RACEWAYS, SUPPORTS, AND OPEN WIRING

The remainder of **Chapter 3** is divided into six groups of wiring methods. The first group is cable assemblies. A cable assembly is a group of conductors under a sheath, armor, or other protective means. For example, Type AC cable consists of insulated conductors wrapped in paper with a flexible corrugated-metal outer sheath (or armor) of aluminum or steel that includes an internal bonding strip of copper or aluminum in intimate contact with the armor for its entire length.

The 11 cable assemblies of **Chapter 3** are listed in alphabetical order. **See Figure 3-9.**

The second group of wiring methods discussed in **Chapter 3** includes 12 different types of circular raceways. **See Figure 3-10.** Raceways, unlike cable assemblies, do not come prewired from the manufacturer, and must have conductors installed after their installation. Raceways are also considered wiring methods. For example, electrical metallic tubing (Type EMT) with insulated type THHN conductors installed is considered a wiring method.

Raceways are circular or rectangular in cross-section and can be metal or nonmetallic. There are two types of tubing: rigid and flexible. Circular raceways can be flexible or rigid type conduits.

The third group of wiring methods, factory-assembled power distribution systems, does not fit into a raceway or cable assembly category and is in a group by itself. The *NEC* recognizes two types of factory-assembled power distribution systems as acceptable wiring methods. These two methods, busway and

Figure 3-9 Group 1, Cable Assemblies

Article	Title of Article
320	Armored Cable: Type AC
322	Flat Cable Assemblies: Type FC
324	Flat Conductor Cable: Type FCC
326	Integrated Gas Spacer Cable: Type IGS
330	Metal-Clad Cable: Type MC
332	Mineral-Insulated, Metal-Sheathed Cable: Type MI
334	Nonmetallic-Sheathed Cable: Types NM and NMC
336	Power and Control Tray Cable: Type TC
337	Type P Cable
338	Service-Entrance Cable: Types SE and USE
340	Underground Feeder and Branch-Circuit Cable: Type UF

***Figure 3-9. Articles 320** through **340** cover the different kinds of cable assemblies found in the* NEC.

Figure 3-10 Group 2, Circular Raceways

Article	Title of Article
342	Intermediate Metal Conduit: Type IMC
344	Rigid Metal Conduit: Type RMC
348	Flexible Metal Conduit: Type FMC
350	Liquidtight Flexible Metal Conduit: Type LFMC
352	Rigid Polyvinyl Chloride Conduit: Type PVC
353	High Density Polyethylene Conduit: Type HDPE Conduit
354	Nonmetallic Underground Conduit with Conductors: Type NUCC
355	Reinforced Thermosetting Resin Conduit: Type RTRC
356	Liquidtight Flexible Nonmetallic Conduit: Type LFNC
358	Electrical Metallic Tubing: Type EMT
360	Flexible Metallic Tubing: Type FMT
362	Electrical Nonmetallic Tubing: Type ENT

Figure 3-10. *The raceways in* ***Articles 342*** *through* ***362*** *have a round cross-section.*

cablebus, are preassembled and bolted together for a complete installation in the field. The result is a grounded, completely enclosed, ventilated, and protective metal housing containing factory-mounted bare or insulated conductors, which are usually copper or aluminum bars, rods, or tubes.

Busways offer a convenient solution for power distribution.

Photo courtesy of Kevin Palm

These systems allow for a disconnecting means and overcurrent protection to be installed anywhere along the busway or cablebus, providing an easy means for power distribution. The two articles for busway and cablebus, **Articles 368** and **370** respectively, separate "Circular Raceways" from "Other than Circular Raceways." **See Figure 3-11.**

There are ten types of raceways that are not circular in cross-section, most

Figure 3-11 Group 3, Factory-Assembled Systems

Article	Title of Article
368	Busways
370	Cablebus

Figure 3-11. *Busways and cablebus have uniquc manufacturer requirements.*

Figure 3-12 Group 4, Raceways that are Other than Circular

Article	Title of Article
366	Auxiliary Gutters
372	Cellular Concrete Floor Raceways
374	Cellular Metal Floor Raceways
376	Metal Wireways
378	Nonmetallic Wireways
380	Multioutlet Assembly
384	Strut-Type Channel Raceway
386	Surface Metal Raceways
388	Surface Nonmetallic Raceways
390	Underfloor Raceways

Figure 3-12. *With the exception of* ***Article 382****, which covers nonmetallic extensions,* ***Articles 366*** *to* ***390*** *cover raceways that are rectangular in cross-section.*

of which are rectangular. These make up the fourth group of wiring methods. **See Figure 3-12.**

The fifth group of wiring methods is surface-mounted nonmetallic branch circuit extensions. Like Group 3, the fifth group does not fit into a raceway or cable assembly category, and so is in a group by itself. It has only one article, **Article 382**, but it is worth noting. **See Figure 3-13.** This article shows two methods to extend a branch circuit. The first method is primarily limited to use from an existing outlet in a residential or commercial occupancy not more than three floors above grade. It has been used primarily in older electrical installations to allow for the surface mounting of additional receptacle outlets. The second method has provisions to allow a "concealable nonmetallic extension" to be installed on walls or ceilings covered with paneling, tile, paint, joint compound, or a similar material. Since the conductors may be concealed only by paint, the conductors must be GFCI protected.

Figure 3-13 Group 5, Surface-Mounted Nonmetallic Extensions

Article	Title of Article
382	Nonmetallic Extensions

Figure 3-13. *Nonmetallic extensions are most commonly used in residential applications.*

Figure 3-14 Group 6, Support Systems and Open Wiring

Article	Title of Article
392	Cable Trays
393	Low-Voltage Suspended Ceiling Power Distribution Systems
394	Concealed Knob-and-Tube Wiring
396	Messenger-Supported Wiring
398	Open Wiring on Insulators
399	Outdoor Overhead Conductors over 1000 volts

Figure 3-14. *The methods included in the final group are not raceways, and are largely considered support methods of wiring.*

The last group consists of six support methods and open wiring. **See Figure 3-14.** Open wiring has limited use and applications. For example, knob-and-tube wiring is an open wiring method that is not allowed in new installations, but many homes still have this method in service.

Each wiring method has advantages and disadvantages depending on the situation. These considerations can be due to economics, required physical protection, environmental location, building construction, and even factors that are unique to a certain jurisdiction. In some cases, the wiring method used may be based upon what the customer desires, often referred to as the "customer's specifications," as long as it is permissible under the *NEC*.

More than one type of wiring method is often used on an electrical project. For example, a branch circuit feeding an electric sign in the front yard of an elementary school may start from an indoor electrical panel located in the school office. As the wiring method and conductors run above the office ceiling, EMT conduit (**Article 358**), which is permissible for this type of installation, could be used. As the conduit penetrates the outside wall of the building and proceeds underground, it is to be routed under a bus stop canopy for a certain distance where it may be subject to physical damage. Based on the danger of physical damage and the rules in **300.4** that require conductors to be protected, intermediate metal conduit, or IMC (**Article 342**), may be chosen.

After clearing the bus stop space, the IMC could transition to PVC (**Article 352**), commonly utilized for underground work, as it continues underground to the sign location. Because the PVC travels underneath a location where buses regularly stop, a burial depth of 24 inches must be used, according to **Article 300.5**. At the sign, the conduit transitions again from PVC to IMC and continues out of the ground to the sign's connection junction box.

Articles 320 to 340, Cable Assemblies

Articles 320 through **340** cover the first group of wiring methods: cable assemblies.

Article 320, Armored Cable: Type AC

Armored, or Type AC, cable, while similar in appearance to metal clad

The Parallel Numbering System

Certain areas of the *NEC* use a parallel numbering system that makes finding a topic within a context easier. In **Articles 320** to **399**, each article uses identical numbering of parts and section topics to make things easier to find. For example, the "**XXX.30**" section of each article (**320.30**, **342.30**, etc.) covers the requirements for supporting and securing that particular wiring method. In accordance with this system, **Article 358** covers installation of EMT and **Section 358.30** covers the support of EMT.

Knowing the commonly used section numbers and their corresponding topics will help the user of the *NEC* find information quickly and consistently. **See Figure 3-15**.

Figure 3-15 Parallel Numbering System for Articles 320 to 399

Part of Article	Section Number	Topic
Part I, General	1	Scope
	2	Definitions
	3	Other Articles
	6	Listing Requirements
Part II, Installation	10	Uses Permitted
	12	Uses Not Permitted
	14	Installation
	15	Exposed Work
	17	Framing Members
	18	Crossings
	19	Clearances
	20	Size
	22	Number of Conductors
	23	Attics
	24	Bends-How Made
	26	Bends-Number In A Run
	28	Reaming, Threading or Trimming
	30	Securing and Supporting
	31	Single Conductors
	40	Boxes and or Fittings
	41	Floor Coverings
	42	Devices
	56	Splices and Taps
	60	Grounding, Bonding
	80	Ampacity
Part III, Construction Specifications	100	Construction
	101	Corrosion Resistance
	104	Conductors
	108	Equipment Grounding Conductor
	112	Insulation
	116	Conduit/Sheath/Jacket
	120	Marking
	130	Standard Lengths

Figure 3-15. *Knowing the commonly used parallel numbering system can make it easy to find information in* ***Chapter 3***.

The Parallel Numbering System (continued)

Each article does not include all of the section numbers listed. An article covering cable assembly, for example, would have a **Section XXX.80** for ampacity, but an article about conduit would not. **Section XXX.28**, Reaming, Threading or Trimming, is dependent on whether the raceway is metallic or nonmetallic. Metallic raceways are reamed and/or threaded, whereas nonmetallic raceways, such as polyvinyl chloride (PVC), are trimmed.

(Type MC) cable, is constructed differently and has advantages and disadvantages of its own. For example, the armor of AC cable can be used as an equipment grounding conductor.

Type AC cables must be installed with an insulating bushing between the conductors and the cable armor, unless equivalent protection is provided. See **Section 320.40** in the *NEC.*

Article 322, Flat Cable Assembly: Type FC

Flat cable (Type FC) assemblies allow taps to be installed along the special raceway that holds the FC. The C-shaped raceway has one side open to allow access to the three to four 10 AWG conductors within the raceway. This can be thought of as a miniature busway, allowing devices to be connected and supplied along the length of the raceway.

Article 326, Integrated Gas Spacer Cable: Type IGS

Integrated gas spacer cable is used for underground service entrance feeder or branch-circuit conductors in large industrial applications. This cable uses sulfur hexafluoride (SF_6) gas as an insulator, as it has a higher insulating strength than nitrogen or air. While SF_6 is not toxic, it displaces oxygen in the lungs, so long-term exposure is not advised and the use of this type of cable is limited to outdoor locations.

The cable has a polyethylene conduit with aluminum conductors and a paper spacer to maintain the spacing of the conductors from the conduit. The conduit is filled with SF_6 gas and maintained at a pressure of 20 pounds per square inch. Valves are provided to check, maintain, and control pressure.

Figure 3-16 CLX Cable

Figure 3-16. *MC cable is available in a wide variety of configurations for different applications including medium voltage and classified locations.*

Courtesy of Okonite Cable Company

Section 326.120 mandates that the conductors range from 250 kcmils to 4,750 kcmils, providing a pathway for very large electrical loads.

For additional information, visit qr.njatcdb.org Item #5512

Due to the superior dielectric strength of SF_6 gas, it is used in high-voltage switchgear indoors where space is limited. Monitoring equipment is installed for leak detection in these contained spaces.

Article 330, Metal-Clad Cable: Type MC

Metal clad (Type MC) cable is similar to AC cable, but is more flexible and can be easier to install. **See Figure 3-16.** Historically, AC cable was used to provide cable with armor that could be used as an equipment grounding conductor, but modern MC cable is constructed with an integral bonding strip that can also be used as an equipment grounding conductor. See **250.118(10)(b)** and **(c)** for more information.

Type MC cable with a bonding conductor and an insulated grounding conductor is typically called "hospital grade" due to its use in the special

For additional information, visit qr.njatcdb.org Item #5513

occupancies, as addressed in **517.13**. Note that the bonding conductor, which is usually an oversized aluminum conductor, is able to short the inductive coil affect that armor could provide in a fault situation. This would limit the current, avoiding the need for the overcurrent protective device to activate.

Figure 3-17 MI Cable Terminations

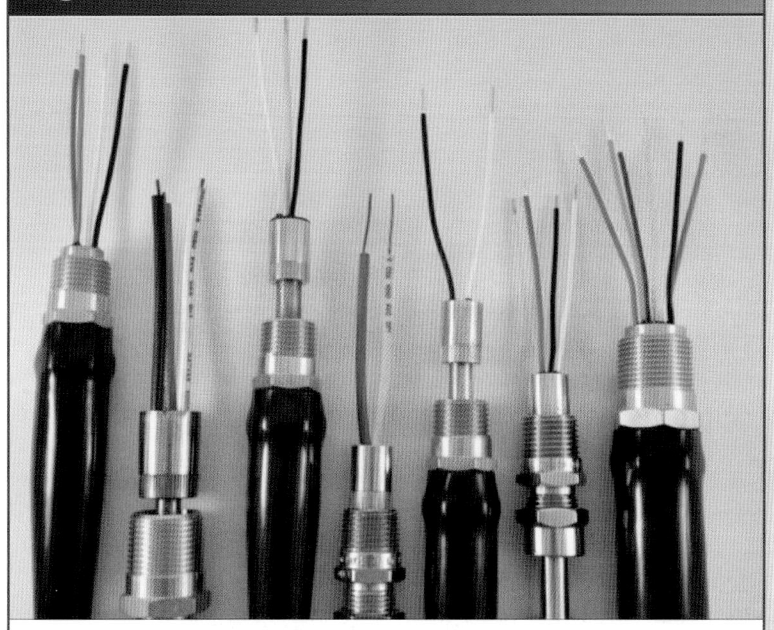

Figure 3-17. *MI cable can be terminated to fit the needs of the application.*
Photo courtesy of MI Cable Company

Figure 3-18 CLX Cable

Figure 3-18. *MI cable is required to be supported every six feet.*
Photo courtesy of Nvent

Article 332, Mineral Insulated, Metal Sheathed Cable: Type MI

Mineral-insulated, metal-sheathed (Type MI) cable offers the advantage that it can be installed almost anywhere and in any conditions when used with the proper fittings. **See Figure 3-17.** As long as the metal sheath is not compromised, it is usually an allowed installation regardless of the location. Type MI cable is available with copper or stainless steel outer sheaths. The conductors within the cable are insulated by a compressed magnesium oxide. **See Figure 3-18.**

Article 334, Nonmetallic Sheathed Cable: Type NM

Nonmetallic sheathed (Type NM) cable originally appeared in the *NEC* in 1926. Initially, it did not have an equipment grounding conductor, but when the *NEC* began to require an equipment grounding conductor for branch circuits in the 1962 edition, NM with grounding conductors began to be used. All NM today has an equipment grounding conductor. More recently-manufactured NM cable also has a color-coded sheath identifying the wire size. **See Figure 3-19.**

This color code is not a requirement of the *NEC*, but is usually done by the manufacturer to improve usability and accuracy of inspection. It is important to note that in much of the existing infrastructure, older Type NM cable is all the same color, so care must be taken when modifying existing circuits.

The conductors of type NM cables have a 90°C rating. The ampacity section of **Article 334** requires that the 60°C ampacity not be exceeded.

There are two types of NM cable: Type NM and Type NMC. Type NMC cable has the same physical attributes as NM cable, but is also resistant to fungus and corrosion. As a result, NMC cable is permitted to be used in moist and damp locations. See **334.10(A)** and **(B)** in the *NEC* for more information.

Article 336, Power and Control Tray Cable: Type TC
Power and control tray (Type TC) cable can be used to supply or control most kinds of circuits covered in the *NEC*. TC cable with the same crush resistance as MC cable is identified as TC-ER. Type TC cable marked TC-JP is suitable for installation in one- and two-family dwelling units when installed according to the relevant parts of **Article 334**. The "-JP" designation indicates that the cable is designed for pulling though framing members such as joists or studs.

Article 337, Type P Cable
Type P cable is new to the 2020 *NEC*. Type P cable is available in armored or unarmored configurations, and has been used for over 40 years in extremely harsh environments such as offshore and land-based drilling operations in the oil and gas industries. These environments involve exposure to severe temperatures, vibration, and corrosion in addition to exposure to a variety of chemicals, abrasives, and oil-based products. **See Figure 3-20.**

Article 338, Service-Entrance Cable: Types SE and USE
Service entrance (Type SE) cable is used for service entrance conductors, as implied by the name. Type USE is for underground use. This type of cable may also be used for branch circuits or feeders. When Type SE cable used for interior wiring, the installer must follow the installation requirements of **Article 334**, excluding **334.80**. Where there is an uninsulated conductor in the cable assembly for a new installation, the uninsulated conductor must be used as an equipment grounding conductor only.

Existing installations may use the uninsulated conductor as a grounded (typically neutral) conductor. This was an allowable use prior to the 1996 edition of the *NEC* for ranges and dryers.

Figure 3-19 Typical Nonmetallic Sheath Colors

Color	Wire Size
White	#14 copper
Yellow	#12 copper
Orange	#10 copper
Black	#6 or #8 copper
Gray	Underground use (Type UF)

Figure 3-19. *While color coding of nonmetallic sheathed cable is not required, manufacturers frequently use different colors to identify the sizes of conductors within the cable assembly.*

Figure 3-20 Type P Cable

Figure 3-20. *Type P cable comes in a variety of configurations for higher voltages, branch-circuit wiring, and control wiring.*

Courtesy of Topher Edwards

Article 340, Underground Feeder and Branch Circuit Cable: Type UF
Underground feeder and branch circuit (Type UF) cable is for underground use and is moisture-, fungus-, and corrosion-resistant. The gray sheath of the cable is bonded around the insulated conductors within the cable. It is not

allowed where subject to physical damage, and appropriate measures must be used to protect the cables when underground and emerging from grade.

Articles 342 to 362: Round Raceways

The second group of wiring methods, round or circular raceways, is covered in **Articles 342** through **362**.

Figure 3-21 Rigid Metal Conduit

Figure 3-21. *Rigid metal conduit (RMC) is used in industrial applications and in any application where physical protection of the conductors is a paramount concern.*

Articles 342 and 344, Intermediate and Rigid Metal Conduits: Type IMC and RMC

Articles 342 and **344** cover the requirements for intermediate metal conduit (IMC) and rigid metal conduit (RMC), respectively. Both of these conduits are used where the conductors may be exposed to physical damage and are permitted to be used in all atmospheric conditions and occupancies, although corrosion protection must still be considered. **See Figure 3-21.** In addition to steel and stainless steel, RMC is available in red brass and aluminum. Where corrosion is of particular concern, PVC-coated variants are available.

RMC has a thicker wall than IMC. IMC is an intermediate thickness, between EMT and RMC. IMC and RMC may both be threaded. Since IMC has thinner walls, it weighs about 67% the weight of RMC. Both types of raceways are considered interchangeable in the *NEC*, though it could be argued that RMC may provide better protection where subject to extreme physical damage. Both kinds of conduit come with color-coded thread protectors to aid the installer in identifying the trade size.

It is important for the installer to note that the support requirements for these raceways vary depending upon size, orientation, and coupling type. These raceways may have their supports spaced up to 20 feet apart.

Article 348, Flexible Metal Conduit: Type FMC

Flexible metal conduit (Type FMC) is commonly used in tenant improvement, remodels, and new construction of office and commercial occupancies. It has limited use as an equipment grounding conductor and may not be used where the total amount of FMC in the entire path exceeds six feet or the circuit exceeds 20 amperes. Another important characteristic of flexible metal conduit is that it may not be used for an equipment grounding conductor in trade sizes over 1 1/4 inches.

Table 348.22 details 3/8-inch FMC requirements. This small-diameter FMC is similar to a metal-clad cable without the conductors installed. The fittings used on this 3/8-inch raceway will influence the number of conductors allowed within. In this unique case, equipment grounding conductors do not count toward raceway fill as they do in other applications.

Until recently, FMC was allowed in wet locations when used with the appropriately listed conductors. However, for current applications in wet locations that also require flexibility and grounding, liquidtight flexible metal conduit is a more appropriate choice.

Article 350, Liquidtight Flexible Metal Conduit: Type LFMC

In addition to being used in wet locations, liquidtight flexible metal conduit (Type LFMC) is allowed in some of the hazardous locations discussed in **Chapter 5** of the *NEC*. LFMC can be used as an equipment grounding conductor per **250.118(6)**. For trade sizes up to 1/2 inch, it can be used for grounding on circuits up to 20 amperes; 60 amperes is the maximum allowed for trade sizes 3/4 inch through 1 1/4 inches. As with all flexible metal raceways, if flexibility is required, a separate equipment grounding conductor must be installed.

Articles 352 through 356: Nonmetallic Conduits

Over the past several *Code* cycles, newer nonmetallic rigid conduits have been added to the *NEC*. Each of these conduits has specific advantages and uses. Polyvinyl chloride (PVC), discussed in **Article 352**, and reinforced thermosetting resin conduit (RTRC), discussed in **Article 355**, are considered rigid non-flexible raceways. The support requirements for these raceways vary by trade size. Temperature changes are also a concern for long runs, but the stable temperature underground is suitable for these conduits.

In **Sections 352.44** and **355.44**, the requirements for expansion fittings are presented. Where these raceways are installed and subject to changes in temperature, such as in outdoor locations, the installer must anticipate the amount of expansion or contraction using the tables in these respective sections. Properly sized expansion fittings must be installed.

Article 353 covers high-density polyethylene (Type HDPE) conduit, a smooth-wall raceway that is resistant to chemicals and provides significant physical protection. It is well suited for underground locations, encasement in concrete, and installation in areas subject to corrosion. **See Figure 3-22.** Special splicing methods, such as heat fusion and electrofusion, or mechanical fittings must be used. See **Article 353** for other requirements.

Nonmetallic underground conduit with conductors (Type NUCC), covered in **Article 354**, is a smooth-wall nonmetallic raceway with conductors pre-installed by the manufacturer. This is an HDPE product with conductors

Figure 3-22 High Density Polyethylene Conduit

Figure 3-22. *HDPE can be used in underground applications, such as supplying power to streetlights.*

already installed. Many similarities exist between NUCC and HDPE, but it is important to note that NUCC has bending radius requirements, which must be adhered to due to the internal conductors already present.

Article 356 covers liquidtight flexible nonmetallic conduit (LFNC). This type of conduit is frequently used where a short run of conductors is needed, such as the distance between a disconnect and an air-conditioning unit for a residence. Longer runs are permitted if supported according to **Section 356.30**.

Article 358, Electrical Metallic Tubing: Type EMT

Electrical metallic tubing (EMT) has been a staple of the electrical industry since the early 1900s. Commonly called "thin-wall," it is a popular choice where flexibility is not required and the environment does not warrant additional physical protection. EMT is not threaded, and instead uses set screw or compression couplings and connectors. With appropriate fittings, it may be used in wet locations.

EMT is available in stainless steel or aluminum in addition to the standard steel material. These special kinds of EMT afford corrosion protection when installed according to **358.10(B)**. Note that the 2020 *NEC* has expanded the use of EMT and should be consulted even if the installer has had extensive prior experience with the product. **See Figure 3-23.**

Figure 3-23 Electrical Metallic Tubing (Type EMT)

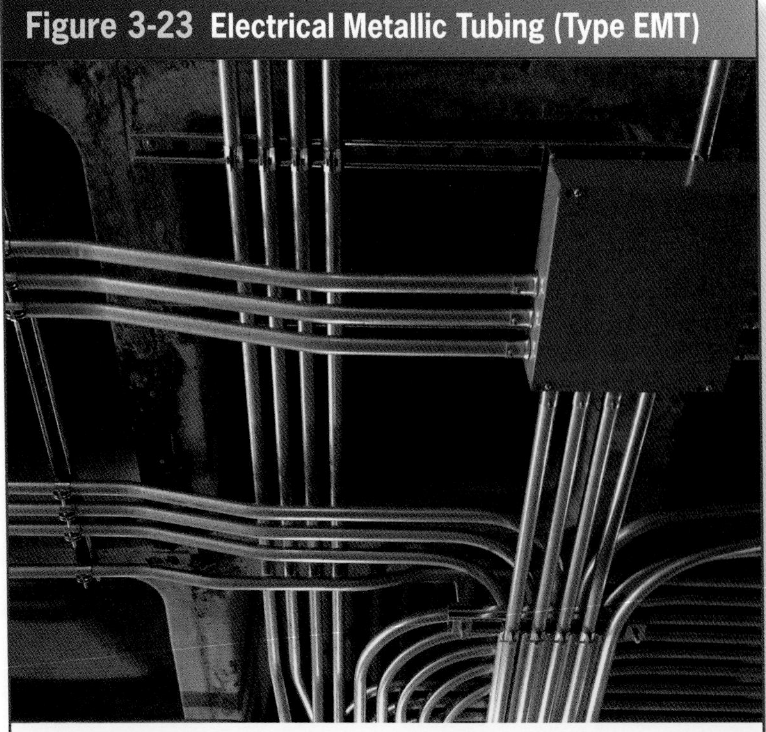

Figure 3-23. *EMT is commonly used in all occupancies to route conductors to their destinations.*

Photo courtesy of Ethan Miller

Article 360, Flexible Metallic Tubing: Type FMT

Flexible metallic tubing (Type FMT) is a thin-wall flexible tubing that can provide a smoke-tight, liquidtight connection. This tubing is only manufactured in 3/8-, 1/2-, and 3/4-inch trade sizes and cannot be used in lengths over six feet. It may be installed in ducts, according to **300.22(B)**. FMC may be installed in this situation as well, but is limited to four feet.

Article 362, Electrical Nonmetallic Tubing: Type ENT

While the *NEC* permits electrical nonmetallic tubing (ENT) to be used in a variety of applications, one of the more common uses is within the concrete floor of a building. The ENT is laid out and secured to the rebar or other concrete structural members, and after the concrete is poured, the conductors are installed. Care must be taken to assure the ENT is not damaged during this process. As with many other raceways, the radius of bends is a concern; these requirements are covered in **Table 2** of **Chapter 9**.

Articles 366 to 390: Rectangular Raceways, Power Distribution Systems, and Nonmetallic Extensions

Group 3, factory-assembled power distribution systems, is covered in **Articles 368** and **370**. Group 4, rectangular (or non-circular) raceways are covered in **Articles 366**, **372** through **380**, and **384** through **390**. **Article 382** covers Group 5, nonmetallic extensions.

Articles 366, 376, and 378, Auxiliary Gutters, Metal Wireways, and Nonmetallic Wireways

Auxiliary gutters and wireways are similar to each other, as they enclose the conductors and perform the same function with the same advantages. The labeling and identification on the product is important and tells the installer which article is to be used when installing the product. If the product is labeled as a wireway, then **Article 376** or **378** is to be used depending on the material of the wireway. If the product is labeled as a gutter, then **Article 366** would be used. In the field, however, workers tend to use the terms *gutter* and *wireway* interchangeably, so care must be taken to discern the actual application and ensure the proper requirements are followed.

An auxiliary gutter, according to **Article 366**, is used to supplement the wiring space for equipment such as panels, meters, and switchboards. Gutters must not be used as a raceway elsewhere in an installation. **Article 366** sets 30 feet as the maximum distance a gutter can be spaced from electrical equipment and still be considered a gutter.

All wireways and gutters use a 20% fill for conductors. Since these items are rectangular in cross-section, this means that these pathways may hold many conductors. The removable or hinged covers provide ready access. Additionally, metal gutters and wireways do not require derating until the number of current-carrying conductors exceeds 30, rather than three as mentioned in **Section 310.16**.

Article 368, Busways

Busways are manufactured with bus bars within them and assembled on site to distribute power throughout a facility. They are able to be tapped with switches and overcurrent devices to supply power to the area as the busway passes through. According to **Section 368.30**, busways must be supported every five feet, and may be used for service, feeder, and branch-circuit conductors.

Article 370, C[illegible]

Cablebus ha[illegible] *NEC* since [illegible] system that [illegible] gle insula[illegible] within a ve[illegible] is capable c[illegible] as 6,000 amperes a[illegible] volts. The enclosures are venti[illegible] keep the conductors within their temperature limits. Once the enclosure is installed, the conductors can be installed within it. Cablebus can be tapped at intermediate points along the run with the proper manufacturer-approved devices.

Articles 372, 374, and 390, Cellular Concrete Floor, Cellular Metal Floor Raceways, and Underfloor Raceways

Articles 372 and **374** include floor systems constructed of concrete or metal that house the conductors. Cellular metal floor raceways are covered with concrete and can be accessed throughout the life of the building. Cellular concrete floor raceways are constructed of precast concrete. With either system, conductors are pulled into the raceways from headers perpendicular to the raceways in the floor.

These systems are installed where access from the floor is needed and future flexibility is desired, such as in office spaces, casinos, administrative areas, and call centers. The headers may have removable covers at finish floor level, or risers can be provided for access. Conductors larger than 1/0 AWG are not generally allowed. Another common point to note is that when outlets are removed, the wiring associated with the outlet must be removed also. If this is not done, the remaining unsecured conductors can tangle up a wire pull, making removal difficult, if not impossible.

Underfloor raceways in **Article 390** are similar to cellular raceways, but the channels are installed such that the cover is flush with the concrete and then covered with carpeting, linoleum, or another floor covering. The restriction on discontinued outlets applies, but this raceway may accommodate any size conductor for which it is designed. See **Section 390.5** for more information.

Article 380, Multioutlet Assemblies

A multioutlet assembly is a wireway or surface raceway with outlets as part of the assembly. This offers a convenient method of providing receptacles where easy access is required, such as along a workbench.

Article 382, Nonmetallic Extensions

Nonmetallic extensions are used to extend branch circuits with flat cabling that may or may not be concealed by the building finish. The device connected to the existing branch circuit outlet must provide GFCI protection. Nonmetallic extensions are limited to use within the room of origin, and voltage between the conductors must be 150 volts or less.

Article 384, Strut Type Raceway

Strut, which is commonly used to build support structures for equipment, can be used as a raceway when properly listed. This offers a convenient raceway and support system for luminaires and other equipment. A metal or nonmetallic cover is installed in the open side of the strut to protect the conductors within, and various fittings are manufactured to access the conductors and supply utilization equipment or to connect other raceways. **Table 384.22** details the cross-sectional areas to be used when calculating the allowable number of conductors in the raceway. The note at the bottom of the table points out that the area can be modified by the coupling or joining means used in the installation.

Articles 386 and 388, Surface Metal and Nonmetallic Raceways

Surface-mounted raceways are rectangular in cross-section and used for power, data, and communication applications. Where different classes of circuits exist in the same raceway, a barrier is provided to separate the circuits. A common use of surface-mounted raceway is to extend a branch circuit in a retrofit or remodel situation to supply lighting, receptacles, or other outlets. It is frequently used in applications of 300 volts or less, but if constructed according to **386.12(2)** it may be used for higher voltage systems. Surface raceways are designed to be used with a certain number and size of conductors determined by the manufacturer. Fittings are supplied to make changes in direction convenient and easy to execute.

Articles 392 to 399: Support Systems and Open Wiring

The sixth group of wiring methods, support systems and open wiring, is covered in **Articles 392** through **399**.

Article 392, Cable Trays

Cable trays are not a type of raceway, but a method of support for raceways, cables, conductors, and cable assemblies. **Table 392.10(A)** lists all of the wiring methods that may be installed in a cable tray. Cable tray is used to support data and communication conductors, power and lighting cables, and medium-voltage cables that may be used on circuits up to 35,000 volts. Cable trays can be of metal or nonmetallic construction. **See Figure 3-24.**

The requirements involved in a cable tray installation can vary widely depending on a multitude of variables, including:

- Voltage
- Single- or multi-conductor cable
- Type of cable tray
  - Ladder
  - Ventilated trough
  - Solid-bottom
- Size of conductors
  - All conductors are same size
  - Conductors are different sizes
- Ampacity of conductors
- Grouping of conductors by circuit

It is important to have a thorough understanding of the installation to be able to apply **Article 392** properly.

Article 393, Low-Voltage Suspended Ceiling Power Distribution Systems

Low-voltage suspended ceiling power distribution systems were introduced in the 2014 edition of the *NEC* and provide another method of low-voltage distribution. The circuit used is limited to Class 2 specifications and 30 volts. A powered grid-rail designed to match the appearance of the suspended ceiling from below is connected to lighting fixtures or other equipment. The net effect is a suspended ceiling grid integrated with a Class 2 supply. Low-voltage lighting, such as LED lighting, can be moved throughout the space quickly and conveniently.

Article 394, Concealed Knob-and-Tube Wiring

Knob-and-tube wiring was the dominant wiring method from the 1880s to the early 1940s. With this method, porcelain supports called knobs were used to support and space the conductors along the wooden framing of a building. Where conductors must pass through the framing, porcelain "tubes" are used to protect the conductors as they passed through the framing members. There are no grounding conductors in this type of system. Although this is an obsolete wiring method and is not allowed in new installations, there are many knob-and-tube installations that are still functioning.

The *NEC* prohibits the use of insulation around knob-and-tube wiring. See **394.12(5)** for more information.

Article 396, Messenger-Supported Wiring

A messenger wire is a wirc uscd to support conductors across a span. Saddles or rings are used to attach and support the conductors along the messenger wire. Messenger wires are usually bare steel wires or cables. This wiring method is frequently a factory-assembled product used for overhead service conductors.

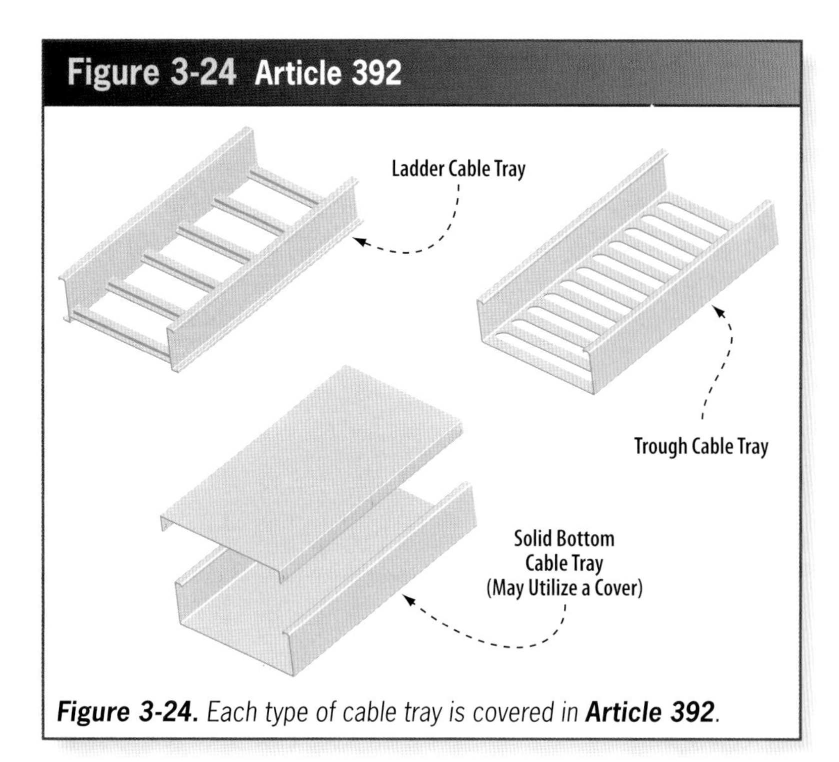

Figure 3-24. *Each type of cable tray is covered in* ***Article 392****.*

Article 398, Open Wiring on Insulators

An open wiring method is not enclosed in a raceway. **Article 398** covers such installations. The conductors are supported on insulators, making this method similar to the knob-and-tube wiring covered in **Article 394**. Open wiring is limited to industrial or agricultural structures. Ways to protect the conductors where exposed to physical damage are listed in **398.15(C)**.

Article 399, Outdoor Overhead Conductors Over 1,000 Volts

Article 399 covers areas where high-voltage conductors are on the load side of a service and run overhead. This method can be used for service conductors, feeders, or even branch circuits if warranted by the equipment demands. These systems must be designed by a licensed professional engineer with experience designing this type of installation.

SUMMARY

The third cornerstone of the *NEC*, **Chapter 3**, shows the requirements for building the installation. Skilled Electrical Workers understand these requirements, enabling them to efficiently install almost any system type. **Chapter 3** covers the entire electrical distribution system, from the service to the last receptacle or outlet in or on the premises. All of the wiring methods used for premises wiring are included in **Chapter 3**.

The chapter is organized first by general requirements, conductors, and enclosures, and then by wiring methods. The parallel number system within **Articles 320** to **399** helps the user of the *NEC* to find a particular requirement quickly.

Chapter 3 is used to build the installation, and the rest of the *NEC* is built upon **Chapter 3**.

QUIZ

Answer the questions below using a copy of the *NEC* and by reviewing Chapter 3 of this text.

1. **A standard four-inch square metal box that is 1 ½ inches deep has a minimum volume of __?__.**
   a. 12.5 in^3
   b. 18 in^3
   c. 21 in^3
   d. 25.5 in^3

2. **Where structural members do not provide ready fastening for EMT, the unbroken lengths of conduit shall be securely fastened within __?__ of structural members.**
   a. 2'
   b. 3'
   c. 4'
   d. 5'

3. **The total cross-sectional area of conductors at any point in a metal wireway shall not exceed __?__ of the interior area of the wireway.**
   a. 10%
   b. 20%
   c. 31%
   d. 40%

4. **The amount of current an 8 AWG THW conductor can carry continuously when installed, according to Section 310.16, is __?__.**
   a. 40 A
   b. 50 A
   c. 60 A
   d. 62.5 A

5. **IMC installed under an airport runway must have a cover depth of __?__.**
   a. 12"
   b. 18"
   c. 24"
   d. 36"

6. **A raceway is installed where it passes from the exterior to the interior of a building and where condensation is a problem. This raceway shall be __?__.**
   a. not made of metal
   b. sealed to prevent warm air from circulating to the cold section of raceway
   c. supported every 3'
   d. none of the above

7. **A medium-voltage cable is rated for up to __?__.**
   a. 2,001 V
   b. 5,000 V
   c. 15,000 V
   d. 35,000 V

8. **When conductors are installed in parallel, they must generally be size __?__ and larger.**
   a. 1/0 AWG
   b. 1 AWG
   c. 3 AWG
   d. 8 AWG

9. **A 1 5/8-inch by 1-inch strut-type channel raceway will have a cross-sectional area of __?__.**
   a. 0.460 in^2
   b. 0.572 in^2
   c. 0.743 in^2
   d. 1.151 in^2

10. **When a cabinet is installed in a wall constructed of wood, the front of the cabinet shall be __?__.**
    a. mounted flush with the surface
    b. mounted such that it is not more than ¼ inch set back from the surface
    c. mounted such that it is not more than ¼ inch in front of the surface
    d. any of the above are permitted

Using the Electricity

Chapter 4 of the *NEC* provides rules and information on electrical equipment for general use. Any "special equipment" that requires unique installation rules is addressed in **Chapter 6**, in accordance with **Section 90.3**.

Chapter 2 of the *NEC* is used for planning the installation of electrical systems and devices and provides information regarding wiring systems and protection. **Chapter 3** provides information for building an electrical installation by delivering electrical energy from the source to the load(s). All equipment covered in **Chapter 4** is dedicated to using the electrical energy. The chapter therefore addresses the control of electrical energy through devices or the consumption of electrical energy by utilization equipment.

Electrical devices are used to provide power to cord- and plug-connected equipment through receptacles, switch lighting, and other loads, and to control other types of electrical equipment that use electrical energy, perform a task, or provide a service for the consumer. For example, motors are used in several different applications, electrical space heaters provide heat, and air conditioners cool homes. All of these devices are covered in **Chapter 4**.

Objectives

- Recall the titles of the articles found in **Chapter 4** of the *NEC*.
- Locate definitions applicable to **Chapter 4** of the *NEC*.
- Apply articles, parts, and tables to identify specific *NEC* **Chapter 4** requirements.

Chapter

Table of Contents

CHAPTER 4 OF THE *NEC*

With the exception of **Articles 400** and **402**, the articles of **Chapter 4** cover pieces of equipment and devices that use electricity. While the intent of transformers (**Article 450**) and phase converters (**Article 455**) is not technically to *use* electricity, each would be seen as a load on the system and both are therefore included in **Chapter 4**. A transformer efficiently changes one AC voltage to another AC voltage, while a phase converter will convert single-phase electrical power to 3-phase electrical power.

Articles 400 and 402, Flexible Cords and Fixture Wiring

Cords and cables are used to connect equipment, and for this reason they are included in **Chapter 4**. **Article 400, Flexible Cords and Flexible Cables**, covers the requirements for these items. Flexible cords are used where flexibility is required, but they may not be used as a substitute for branch-circuit wiring. For branch circuits that need flexibility, metal-clad (MC) cable with stranded conductors or flexible metal conduit (FMC) are good choices.

Section 400.10 shows what uses are permitted for cords and cables, including:

- Connecting pendants
- Wiring of luminaries (lighting fixtures)
- Elevator cables
- Connecting equipment to facilitate frequent changes
- Connection of moving parts

Flexible cords are also often used with attachment plugs for equipment energized by a receptacle.

According to **Section 400.12**, cords and cables cannot be:

- Used in the fixed wiring of a structure
- Run through walls or in suspended ceilings
- Run through doorways or windows
- Attached to surfaces of buildings

Conductors within a cord or cable, as with any other conductor, have ampacity restrictions. These ampacity values are listed in **Tables 400.5(A)(1)** and **(A)(2)**. It is important to note that the ampacity may vary according to the number of current-carrying conductors present in the cord or cable. Where there are more than three current-carrying conductors in a cable, the ampacity is reduced to the percentage factor given in **Table 400.5(A)(3)**.

Similar to **Table 310.4**, **Table 400.4** contains the construction specifications for various cords and cables. Different cables are able to withstand different environmental conditions and are able to resist varying levels of physical damage depending on their construction. The size and number of the conductors available are also shown in the table. Pay attention to the notes at the end of the table, which hold other relevant information that cannot fit within the table itself.

Article 402 provides information and requirements for the use and limits of fixture wire in luminaires and associated equipment. It is important to note that fixture wires are not permitted to serve as branch circuits, but are permitted for installation in lighting fixtures or associated equipment. As covered in **Table 402.3**, fixture wires are rated at higher temperatures than general use conductors due to the heat produced in most luminaire fixtures. Fixture wires are also allowed for Class 1 circuits (see **Article 725**) and fire alarm circuits (see **Article 760**). **Table 402.3** contains the construction specifications and rating of fixture wires.

Articles 404 and 406, Switches and Receptacles

While switches and receptacles do not use electricity themselves, they are involved in the control and use of electricity, which justifies their inclusion in **Chapter 4**.

Article 404, Switches, also covers circuit breakers when they are used as switches. Circuit breakers arranged to be used as switches must be rated accordingly. See **240.83(D)** in the *NEC*.

Fact

Tamper resistant receptacles are required where young children are likely to be present and may insert conductive objects such as keys into a receptacle. Weather resistant receptacles are required in damp or wet locations, regardless of the kind of enclosure or cover used.

For example, a commercial occupancy may have its lights on 24 hours a day, but the lights only need to be illuminated between the time when people arrive in the morning and the time when they leave in the evening. The circuit breakers could be used as a control in this instance, but they would need to be rated for switching due to the frequent use.

In general, switches must be readily accessible, and according to **Section 404.8**, must be installed such that the user does not have to reach higher than six feet and seven inches to operate the switch when it is in its highest position. Other codes and standards may influence mounting heights, including requirements within the Americans with Disabilities Act (ADA). **See Figure 4-1.** See **Annex J** in the *NEC* for more information about these guidelines.

When an enclosure containing switches has voltage between the conductors greater than 300 volts, a barrier must be inserted between the switches. This requirement also applies to receptacles, covered in the next article.

Article 406 covers receptacles, cord connectors, and cord caps or attachment plugs. As with switches, the information in **Annex J** may prove useful for mounting heights depending on the occupancy. Receptacles have the same barrier requirement as switches when more than 300 volts are present in the enclosure between adjacent devices. See **406.5(J)**.

Over the past few *Code* cycles, several types of receptacles have been added to **Article 406**. There are now requirements for tamper resistant (TR) receptacles and weather resistant (WR) receptacles.

Other receptacles for countertop or work surface use and receptacles with USB chargers are also included in **Article 406**. **See Figure 4-2.** Generally, while the requirements in the current edition of the *NEC* will apply to all new installations, the *Code* does not require existing installations to be brought up to the current standards; however, this is not the case with

Figure 4-1 Figure J.6 From Annex J, ADA Standards

J.6 Forward Reach.

J.6.1 Unobstructed. Where a forward reach is unobstructed, the high forward reach shall be 1220 mm (48 in.) maximum, and the low forward reach shall be 380 mm (15 in.) minimum above the finish floor or ground. (See Figure J.6.1.)

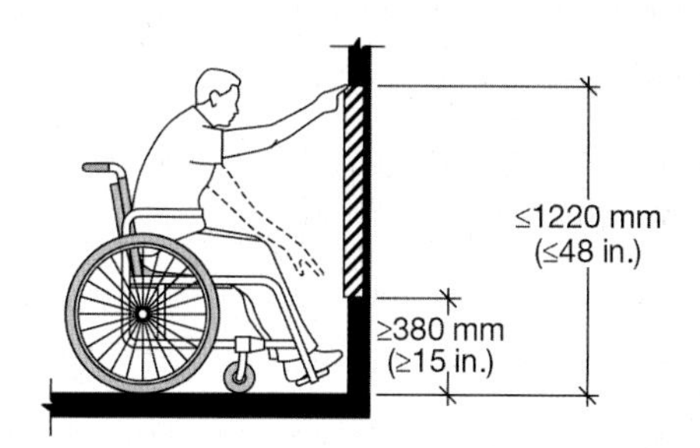

FIGURE J.6.1 Unobstructed Forward Reach.

***Figure 4-1. Informative Annex J** provides guidance for locating devices where ADA requirements apply.*

Figure 4-2 Duplex Receptacle with USB Chargers

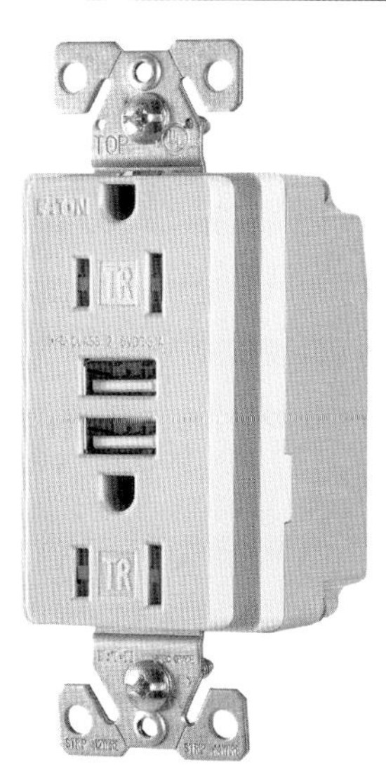

***Figure 4-2.** NEC **406.3(F)** addresses receptacles with integrated USB chargers.*

Photo courtesy of Eaton

replacement of receptacles. Where a receptacle is replaced in a location that requires a GFCI, AFCI, TR, or WR receptacle, it must be replaced with the kind of receptacle currently required by the *NEC*. See **406.4(D)**; note that exceptions may apply.

Article 408, Switchboards, Switchgear, and Panelboards

The articles in **Chapter 4** include provisions for the control and protection of utilization equipment. In most cases, this occurs at the source of the feeder(s) and branch circuit(s) supplying the equipment. Switchboards typically provide distribution to larger loads and other panels. **See Figures 4-3 and 4-4.**

> **REMINDER**
>
> It is important to be aware of the proper definitions of panelboard and switchboard. What many consider an electrical panel is actually a panelboard within an enclosure. These enclosures frequently have hinged covers, which makes them cabinets by definition. See **Article 100, Definitions.**

Figure 4-3 Switchboard

Figure 4-3. *Switchboards typically supply service to larger loads such as motors, transformers, and larger-ampacity equipment.*

Article 410, Luminaires, Lampholders, and Lamps, and Article 411, Low-Voltage Lighting

Article 410 covers luminaires, lighting fixtures, lighting equipment of various kinds, and the wiring and equipment of such products and their installation. As of the 2020 edition of the *NEC*, there are 16 parts to **Article 410**, including the new **Part XVI, Special Provisions for Horticultural Lighting Equipment**. **Part II**, which is used most frequently by installers, contains the requirements for luminaires in various locations. **Sections 410.10** through **410.18** list the requirements for 13 different locations. **Part IV** details the requirements for the support of luminaires, ranging from outlet boxes to structures or other supports, even including trees.

Article 410 follows the same pattern laid out elsewhere in the *NEC*, with sections becoming more specific as the article progresses. This includes sections on particular lighting fixtures such as track lighting and recessed luminaires, commonly called "can lights."

Article 411, Low-Voltage Lighting, first appeared in the 1996 *NEC*. At the time, low-voltage lighting was being used more and more to provide illumination for task and general lighting in office and retail locations. The conductors feeding these luminaries are frequently exposed conductors hung from the ceiling, and can be run the length of a room. Light fixtures are connected to these conductors and can be easily located anywhere along the run, whether to highlight specific merchandise or to modify lighting in a work area.

The power supply for a low-voltage lighting circuit is limited to 30 volts AC, or 60 volts DC, and 25 amperes. If wet contact with the lighting is likely to occur, as may be the case in some landscape lighting installations, the voltage is limited to 15 volts AC or 30 volts DC.

Article 422, Appliances

Upon seeing the term *appliance*, a first-time user of the *Code* may imagine only large kitchen appliances, but a quick glance through **Article 422** shows that this term means much more. The definition for *appliance* from **Article 100** is:

Appliance. Utilization equipment, generally other than industrial, that is normally built in standardized sizes or types and is installed or connected as a unit to perform one or more functions such as clothes washing, air-conditioning, food mixing, deep frying, and so forth.

Figure 4-4 Panelboard

Figure 4-4*. Panelboards are typically used for branch circuits to supply wiring devices such as receptacles and luminaires.*

Article 422 covers electrical appliances regardless of the occupancy in which they are used. However, while the definition of *appliance* technically includes air-conditioning equipment, this equipment is covered separately in **Article 440, Air-Conditioning and Refrigerating Equipment**. It is also important to consider that many electrical appliances are motor-operated, in which case the requirements in **Article 430** must be followed.

Section 422.5 lists those appliances that require GFCI protection. This provides a convenient way to determine whether an appliance being installed requires GFCI protection by one of the means covered in **422.5(B)**. Note that this does not negate other GFCI requirements included in the *NEC*, and **Section 210.8** is specifically referenced here.

Below is a list of appliances that have specific requirements included in **Article 422**:

- Central heating equipment
- Storage-type water heaters
- Central vacuum outlets
- Garbage disposers
- Built-in dishwashers
- Built-in trash compactors
- Wall-mounted ovens
- Counter-mounted cooking units
- Range hoods
- Microwave oven / range hood combo units
- Ranges

There are other appliances with requirements in other articles, such as ceiling fan supports in **Article 314**, which **Article 422** directs the user to.

Any piece of equipment will require maintenance or eventual replacement. **Part III** of **Article 422** contains the disconnect requirements for appliances. In many cases, smaller appliances shall be permitted to use the cord and plug to disconnect. In others, the disconnecting means must generally be either in sight or capable of being locked in the open position.

In Sight From (Within Sight From, Within Sight). Where this *Code* specifies that one equipment shall be "in sight from," "within sight from," or "within sight of," and so forth, another equipment, the specified equipment is to be visible and not more than 15 m (50 ft) distant from the other.

Articles 424 to 427, Heating For Spaces, Processes, Pipelines, and Snow-Melting Equipment

These four articles cover types of heating that, with the exception of space heating in **Article 424**, are primarily used in industrial or commercial occupancies. **See Figure 4-5.**

Article 430, Motors, Motor Circuits, and Controllers

Motors have such unique electrical characteristics and such a multitude of applications that they have an article to themselves. **Figure 430.1**, like many other figures throughout the *NEC*, can help guide the user to the correct section of the article. **See Figure 4-6.**

When started, motors draw a large amount of current, called *inrush*. Therefore the overcurrent protection requirements for a motor load are different than for a non-motor load, where inrush is limited or non-existent. With a non-motor load, overload protection of the equipment and short-circuit protection of the conductors is provided in one device, such as a fuse or circuit breaker. With a motor, however, since the inrush current is six to ten times the normal running current, sizing a breaker to the running current can result in nuisance tripping of the overcurrent device.

*Requirements for installation of motors are covered in **Article 430.***

Furthermore, motors from different manufacturers have differing amounts of current draw, even though the horsepower of the load does not change. Unless current values used for designing motor circuits are standardized,

Figure 4-5 Scope of Articles 424 to 427

Article	Scope
424 Fixed Electric Space-Heating Equipment	Fixed electric equipment used for space heating. For the purpose of this article, heating equipment includes heating cables, unit heaters, boilers, central heating systems, or other fixed space heating equipment.
425 Fixed Resistance and Electrode Industrial Process Heating Equipment	Fixed industrial process heating using electric resistance or electrode heating technology. Heating equipment shall include boilers, electrode boilers, duct heaters, strip heaters, immersion heaters, processed air heaters, or other fixed electric equipment used for industrial process heating.
426 Fixed Outdoor Electric Deicing and Snow-Melting Equipment	Fixed outdoor electric deicing and snow melting equipment and the installation of these systems.
427 Fixed Electric Heating Equipment for Pipelines and Vessels	Electrically energized heating systems and the installation of these systems used with pipelines and vessels.

Figure 4-5. *The scope of these articles show the types of equipment covered by each article.*

replacement may be a challenge, as sizing the conductors and overcurrent devices based on a specific motor nameplate could result in a need to replace the entire installation whenever the motor needs replacement.

This first problem of standardization is dealt with in **Section 430.6(A)(1)**. In this section, standard tables are used to determine the current used for sizing conductors, short-circuit and ground-fault protection, and disconnecting means for most motors. These tables are found at the end of **Article 430** and are organized according to the type of motor used.

- **Table 430.247** is used to determine the full load amperage (FLA) of DC motors.
- **Table 430.248** is used to determine the FLA of single-phase AC motors.
- **Table 430.249** is used to determine the FLA of 2-phase motors.
- **Table 430.250** is used to determine the FLA of 3-phase motors.

Fact

Generally, 2-phase motors are only used in special applications of historical significance, as these electrical systems were some of the original systems used in the United States. Their phase voltages are 90 electrical degrees apart, rather than 120 electrical degrees as in the current 3-phase systems.

Figure 4-6 Figure 430.1

General, 430.1 through 430.18	Part I
Motor Circuit Conductors, 430.21 through 430.29	Part II
Motor and Branch-Circuit Overload Protection, 430.31 through 430.44	Part III
Motor Branch-Circuit Short-Circuit and Ground-Fault Protection, 430.51 through 430.58	Part IV
Motor Feeder Short-Circuit and Ground-Fault Proteciton, 430.61 through 430.63	Part V
Motor Control Circuits, 430.71 through 430.75	Part VI
Motor Controllers, 430.81 through 430.90	Part VII
Motor Control Centers, 430.92 through 430.99	Part VIII
Disconnecting Means, 430.101 through 430.113	Part IX
Adjustable-Speed Drive Systems, 430.120 through 430.131	Part X
Over 1000 Volts, Nominal, 430.221 through 430.227	Part XI
Protection of Live Parts—All Voltages, 430.231 through 430.233	Part XII
Grounding—All Voltages, 430.241 through 430.246	Part XIII
Tables, Tables 430.247 through 430.252	Part XIV

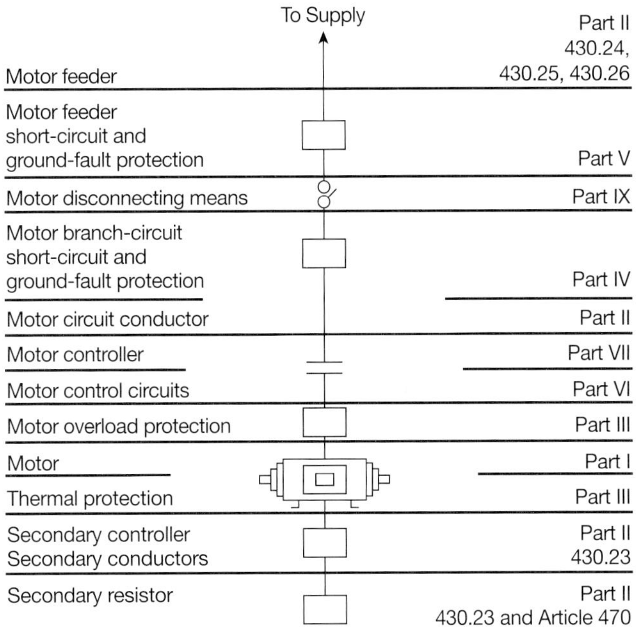

***Figure 4-6. Figure 430.1** is for informational purposes to aid the user in finding the relevant reference.*

Motor Branch-Circuit Conductors

Section 430.22 is used to properly size the conductors for a motor branch circuit. According to **430.22(A)**, for most motors, 125% of the FLA from the tables is used to determine the ampacity of the conductor. Other types of motors and special applications are mentioned in **Section 430.22**. In some cases, as in **430.22(E)**, the motor nameplate full load current is used. These branch-circuit conductors supplying the motor must be protected against overloads, short circuits, and ground faults.

Motor Overload Protection

Part III of **Article 430** has requirements for overload protection. It is important to note the size of motor and kind of overload protection used. For most motors over one horsepower, a separate overload device is used. These kinds of overload devices are usually part of an overload relay, and they sense the amount of current the motor is using either thermally or magnetically. When the current in the circuit is too high for too long, a contact or contacts will open, which shuts off the motor. Since this protection is sized to protect a particular motor, the nameplate current for the motor is used as a reference. See **430.32(A)(1)**. Depending on the service factor or temperature rise on the nameplate, overload protection is sized at 115% or 125% of the motor's nameplate current.

Short-Circuit and Ground-Fault Protection

Article 430.52 in **Part IV** discusses short-circuit and ground-fault (SCGF) protection. *NEC* **240.4(D)** makes it clear that the rules for overcurrent protection of small conductors do not apply to motor circuits, and **240.4(G)** permits **Parts II** through **VII** to be applied. Using the table values in the back of **Article 430** and referencing **430.6(A)(1)** for the full load current of the motor, **Section 430.52** is used to size the SCGF protection.

Suppose an Electrical Worker needs to install a 480-volt, 10-horsepower motor. The nameplate of the motor is 12.4 amperes with a service factor of 1.15. All terminations and conductors in this case are rated at 75°C.

To size the branch-circuit conductors, the worker looks up the full load current in **Table 430.250** and sees that it is 14 amperes. According to **430.22**, 125% of this value shall be used, which is 17.5 amperes. The ampacity **Table 310.16** shows that a #14 AWG conductor will satisfy this requirement.

The overload protection is sized using **430.32(A)(1)**. The nameplate current of the motor is 12.4 amperes, and since the service factor is 1.15, the maximum value of overload protection used shall be 125% of 12.4 amperes, or 15.5 amperes. The overload device used should come as close as possible to 15.5 amperes without exceeding it.

The table value of 14 amperes is used as a reference to size the SCGF protection. Different kinds of overcurrent protective devices have different responses to current, which is one reason **Table 430.52** offers differing percentages based on the overcurrent protective device chosen. In this example, a dual-element time-delay fuse is used. According to **430.52(C)(1)**, the value of the fuse cannot exceed 175% of the full load current of 14 amperes, or 24.5 amperes. However, according to **Section 240.6**, there is no such standard size fuse. An exception is provided in **430.52(C)(1)** that allows the next standard size, in this case a 25-ampere fuse, to be used.

When using the *NEC*, many different articles and sections may be required to solve a problem. Notice how many sections of the *Code* the Electrical Worker must use to size one branch circuit. He or she must be able to find this information in a reasonable amount of time and apply it with confidence. Compliance with the entire *Code* is essential.

Article 440, Air-Conditioning and Refrigeration Equipment

Article 440 is concerned with air conditioning and refrigeration equipment that uses hermetic refrigerant motor compressors. Equipment that does not incorporate a hermetic refrigerant motor compressor must be installed according to the applicable rules of **Articles 422**, **424**, or **430**. See **440.3(B)**. The information in **Article 440** adds to or amends other articles in the *NEC*, including **Article 430**. Similar to **Article 430**, **Article 440** also covers the sizing of branch circuits, overload protection, and SCGF protection for this equipment.

For this equipment, the rated full-load current or branch-circuit selection current on the equipment is used rather than table values. These can be key terms to look for when deciding whether to use **Article 430** or **Article 440**. Most equipment manufacturers will include the conductor size and appropriate overcurrent protection to use on the nameplate, as required in **Article 440**.

Room air conditioners are covered in **Part VII** of **Article 440**. This includes window, wall, or console units. These units are all limited to 250 volts, single-phase.

Article 445, Generators, Article 450, Transformers, and Article 455, Phase Converters

Articles 445, **450**, and **455** involve power generation or altering a power source supply.

Article 445 provides information and requirements for generators, which are considered equipment for general use. In accordance with **Section 90.3**, the *NEC* addresses "special" energy systems in **Chapter 6**. **Article 445** addresses the generator requirements of **Chapter 4** and is not divided into parts.

Transformers are also located in **Chapter 4**. When utilization equipment operates at different system voltages in an electrical installation, transformers are used to derive new systems at the utilization voltage to meet the system requirements. This is an essential part of most electrical installations. Transformers provide an installation with the flexibility of deriving a new system to meet the requirements of utilization equipment. **Article 450** contains extensive requirements concerning the location, overcurrent protection, and other considerations specific to various kinds and sizes of transformers. Furthermore, transformer vaults are covered in **Part III** of this article.

Phase converters, covered in **Article 455**, allow economical and efficient use of utilization equipment. For example, consider an older building or structure with a single-phase, 3-wire service. To use a 3-phase motor in this occupancy, a phase converter would be required. A more modern solution that has become commonplace is to use a variable frequency drive (VFD). Care must be used in selecting a VFD that is capable of this function.

Article 460, Capacitors, and Article 470, Resistors and Reactors

Capacitors are used for power factor correction, for filtering for power supplies, and as add-on parts to motors to improve torque and performance. **Article 460** covers capacitors that are connected to circuits, but does not include those that are part of equipment. When used in motor circuits, they reduce the current needed for the motor by reducing the inductive effects the motor has on the circuit. One important safety note when working with capacitors is to discharge them before handling. **Article 460** contains stipulations regarding the discharge of the capacitors depending on whether the circuit is above 1,000 volts. See **Sections 460.6** for capacitor circuits 1,000 volts or less and **460.28** for capacitor circuits above 1,000 volts.

Resistors and reactors, covered in **Article 470**, are used in motor applications, but are not used to improve power factor. Resistors are used in wound rotor motors to vary the resistance in the rotor circuit, limit the inrush, and control the torque. Resistors are also used in wye-delta starting of motors where the power is not interrupted as the motor controller switches the motor from a wye to a delta configuration.

By reducing the voltage on a motor during starting, inrush currents are reduced and smoother initial acceleration can be provided. Reactors are used with motors for this purpose and are also used on the line and load side of a VFD. These reactors can filter high-frequency currents on the line side of the VFD and protect the VFD on the load side under short-circuit conditions. Reactors can also reduce the voltage spikes that accompany the use of a VFD located a long distance away from the motor it is driving.

Article 480, Storage Batteries

Article 480 applies to all stationary installations of storage batteries. Batteries are commonly used as energy storage systems for photovoltaic (PV) and uninterruptible power supplies. Some facilities, such as main telecommunication centers for phone companies, have multiple floors of a building dedicated to back-up power. **Article 480** would be involved in energy storage applications from this level down to small off-grid stand-alone PV systems.

The sizing of conductors, battery rack construction, disconnecting means, ventilation, and accessibility all need to be considered when installing batteries. These concerns are all addressed in **Article 480**.

Figure 4-7 High Voltage Injury

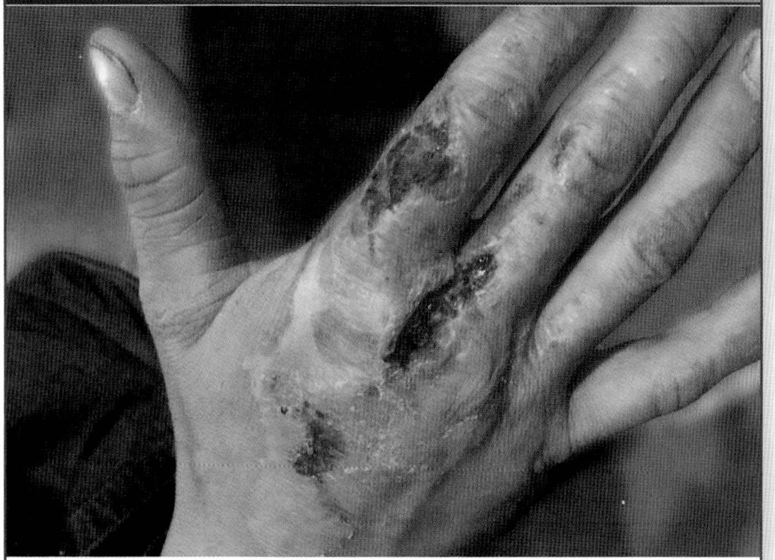

Figure 4-7. *Safety and proper use of testing equipment cannot be emphasized enough when working around high voltages.*

Article 490, Equipment Over 1,000 Volts, Nominal

Many articles in the *NEC* include a section for applications over 1,000 volts. It is important for the user of the *Code* to realize that for each one of these references, **Article 490** may also apply to the particular installation.

Article 490 also includes requirements for substations. Typically substations are under the purview of utilities and outside the scope of the *NEC*, but large industrial or high-technology sites may have substations that are privately owned and not under the control of a utility. While safety is always an important concern, extra precautions must be taken when working on high-voltage circuits. **See Figure 4-7.**

SUMMARY

Chapter 4 of the *NEC* is titled **Equipment for General Use**. This chapter is the final corner of the foundational **Chapters 1** through **4** that apply generally to all installations. The requirements for planning the branch circuits supplying the equipment are found in **Chapter 2**, and wiring methods to build out the installation are covered in **Chapter 3**. The equipment in **Chapter 4** uses the electricity. **Chapter 4** also covers flexible cords, switches, and receptacles that are used to supply the equipment using the electricity. While applying these specific requirements, the worker must also comply with the rest of the *NEC*.

QUIZ

Answer the questions below using a copy of the *NEC* and by reviewing Chapter 4 of this text.

1. **Generally, when sizing the conductors for the branch circuit of a single-phase motor, the information from __?__ is used.**
   a. the motor nameplate
   b. **Table 430.7(b)**
   c. **Table 430.247**
   d. **Table 430.248**

2. **A 125-volt, 15-ampere non-locking receptacle controlled by an automatic control device must be __?__.**
   a. identified by a unique color
   b. marked with a symbol indicating a controlled receptacle
   c. marked with the word "controlled"
   d. both b. and c.

3. **When using capacitors in a power circuit, the conductors connected to the capacitors must have an ampacity of __?__ of the rated current of the capacitor.**
   a. 80%
   b. 100%
   c. 125%
   d. 135%

4. **Electric vehicle cable designed for extra-hard usage has the type letter __?__.**
   a. EVE
   b. EVJ
   c. EVJE
   d. EVJT

5. **For the purpose of sizing the branch circuit of a 100-gallon storage-type water heater, the water heater shall be considered a __?__ load.**
   a. continuous
   b. noncontinuous
   c. normal
   d. supplemental

6. **Switches installed in a tub or shower space __?__.**
   a. may be installed if they are a part of the shower or tub assembly
   b. must be mounted flush with the surface
   c. must have a weatherproof enclosure
   d. shall never be installed

7. **According to the *NEC*, 1.2 volts is a common voltage for a(n) __?__ battery cell.**
   a. alkali
   b. lead-acid
   c. lithium-ion
   d. none of the above

8. **A 4,150-volt electrode-type boiler must not have a control voltage that exceeds __?__.**
   a. 120 V
   b. 150 V
   c. 600 V
   d. 2,075 V

9. **A rotary phase converter __?__.**
   a. converts single-phase current to 3-phase current
   b. has a rotary transformer
   c. has capacitors
   d. all of the above

10. **Conductors supplying luminaires on luminaire chains must __?__.**
    a. be sized 12 AWG minimum
    b. be solid
    c. be stranded
    d. not be larger than 10 AWG

Chapter 5 of the *NEC*: Special Occupancies

This chapter introduces **Chapter 5** of the *NEC*, which contains requirements that are unique to specific occupancies. In other codes and standards, occupancies are categorized according to their general use, such as business, education, or institutional. In the *NEC*, occupancies are categorized according to their specific use and purpose. **Chapter 5** is organized by occupancies and their associated electrical or safety concerns.

Chapter 5 in the *NEC* modifies the other chapters of the *NEC* with the exception of **Chapter 8**, which is a stand-alone chapter. Various occupancies bring with them the concerns of the safe access and egress of large numbers of people, the processing of hazardous materials, the medical care of people, and the housing of livestock. Special structures also addressed in this chapter include manufactured and mobile homes as well as marinas and boatyards. While not strictly an occupancy type, temporary installations can occur in any occupancy, and as such **Article 590, Temporary Installations**, brings **Chapter 5** to a close.

Objectives

- » Recall the titles of the articles found in **Chapter 5** of the *NEC*.
- » Locate definitions applicable to **Chapter 5** of the *NEC*.
- » Identify the requirements of **Article 500** and their relationship to **Articles 501** through **504**.
- » Apply articles, parts, and tables to identify specific *NEC* **Chapter 5** requirements.

Chapter 5

Table of Contents

CHAPTER 5 OF THE *NEC*

Chapter 5 of the *National Electrical Code* is the first of the three special chapters. As stated in **Section 90.3**, the previous chapters of the *NEC* apply generally, and the special chapters (**5**, **6**, and 7) modify or supplement **Chapters 1** through 7.

When using the *Codeology* method of finding information in the *NEC*, it is important to pay attention to the occupancy as it is likely to have a significant impact on the installation. The first articles of **Chapter 5** show this clearly.

Hazardous Locations

Standards pertaining to hazardous locations, also called *classified locations* in the *NEC*, begin with **Article 500** and end in **Article 517**. **Articles 501** through **506** have requirements related to particular hazards, such as flammable vapors, combustible dusts, and fibers or flyings, while **Articles 510** through **517** are specific to the hazardous area or occupancy type.

An important part of a new *Code* user's experience is becoming familiar with hazardous locations. Users will often think of industrial locations as the only hazardous locations, but in reality, the general public typically visits one of the most common hazardous locations at least once a week to refuel their cars. Portions of gas stations used to refuel vehicles fall under the classification of hazardous locations; however, only the area where the gas is being dispensed is considered hazardous. The convenience store associated with the typical gas station is not considered a hazardous location, and is covered by **Chapters 1** through **4** of the *NEC*.

The informational notes in **Section 500.5** of **Article 500** give examples of occupancies where the user may find specific hazardous or classified locations. There are two ways an occupancy can be classified: the *Class and Division* method or the *Zone* method. The Class and Division method has been in the *NEC* since 1923, while classification by Zone is relatively new and was added to the *NEC* in 1996. While the previous method had been in use in the United States for decades, the Zone method allowed for international harmonization when it came to protection from explosions related to the use of electricity. This allows manufacturers and designers across the globe access to one universal standard.

Article 500

Article 500 in the *NEC* has an unusual scope. As stated in **Section 500.1**, **Article 500** must be incorporated with **Articles 501** to **503** (as well as **Article 504**). **Article 500** is a general overview of the requirements related to hazardous locations and is not used by itself. The information contained in **Article 500** must be coupled with the appropriate article addressing the specific classified hazard involved.

Article 501 concerns flammable gases, **Article 502** covers combustible dusts, and **Article 503** contains requirements for areas with fibers or flyings. One way to remember this distinction is by relating the articles to the size of the various hazardous particle, which increases from gas molecules in **Article 501** to fibers in **Article 503**. Note that, as the size of the particle increases, the number of related requirements decreases from **Article 501** to **503**.

Electrical installations in classified areas require electrical materials and equipment designed to prevent sparks from occurring around flammable vapors, liquids, gases, and dusts/fibers. This type of equipment is not used in basic commercial and industrial locations and may be unfamiliar to the installer. Items such as hazardous-rated push-button switches, receptacles, and light fixtures must be installed in classified areas, and the conduit system must also be appropriately hazardous-rated. The conduit system has to prevent ignitable vapors, liquids, gases, and dust/fibers from entering the raceway and contacting the equipment. Junction boxes, flexible connections, and 90-degree pull boxes must all be hazardous-rated.

500.5 Classifications of Locations

The classification of hazardous locations is based upon two factors:

1. The properties of the flammable vapors, liquids, gases or liquid-produced vapors, or combustible dusts or fibers that may be present
2. The likelihood that a flammable or combustible concentration or quantity is present

500.5(B) Class I Locations

Class I locations are those in which flammable gases or vapors produced by flammable liquid are or may be present in the air in quantities sufficient to produce explosive or ignitable mixtures.

Class I locations are further divided into Division 1 and Division 2 locations based upon the immediacy of the hazard posed by flammable gases or vapors.

(1) Class I, Division 1
- Locations in which ignitable concentrations exist under normal operations
- Locations in which ignitable concentrations may exist due to repair, maintenance, or leaks
- Locations in which ignitable concentrations may exist due to processes, breakdown, or faulty equipment

(2) Class I, Division 2
- Locations in which volatiles are handled, processed, or used in closed containers
- Locations in which positive ventilation prevents accumulation of gases or vapors
- Areas adjacent to Class I, Division 1 locations

500.5(C) Class II Locations

Class II locations are those that are hazardous because of the presence of combustible dust. An example of a Class II location is a grain storage bin that produces combustible dust during the movement of the grain. **See Figure 5-1.** Other examples of Class II hazardous locations are flour and feed mills; producers of plastics, medicines, and fireworks; producers of starch or candies; and spice-grinding, sugar, and cocoa plants.

Class II locations are further divided into Division 1 and Division 2 locations based upon the immediacy of the hazard posed by combustible dust.

(1) Class II, Division 1
- Locations in which ignitable concentrations of combustible dust exist under normal operations
- Locations where mechanical or machinery failure, repair, maintenance, or leaks could create ignitable concentrations of dust
- Locations in which Group E metal dusts exist, including but not limited to aluminum and magnesium

Class II, Division 2
- Locations in which ignitable concentrations of combustible dust may exist due to abnormal operations
- Locations in which combustible dust is present, but is not present in ignitable concentrations unless a malfunction of equipment or process occurs
- Locations in which accumulating dust interferes with heat dissipation and/or could be ignited through equipment failure

Figure 5-1 Grain Processing Facility

Figure 5-1. *Precautions must be taken where combustible dusts are present to prevent explosions in locations such as grain processing facilities.*

500.5(D) Class III Locations
Class III locations are those that are hazardous because of the presence of easily ignitable fibers or flyings, but where such fibers or flyings are not likely to be in suspension in the air in quantities sufficient to produce ignitable mixtures.

Class III locations are further divided into Division 1 and Division 2 locations based upon the immediacy of the hazard posed by combustible fibers or flyings.

Figure 5-2 Parallel Numbering of Articles 501 to 503

Section Number	Topic
10	Wiring Methods
15	Sealing and Drainage
17	Process Sealing
20	Conductor Insulation
25	Exposed Parts
100	Transformers and Capacitors
104	Conductors
105	Meters, Instruments, and Relays
115	Switches, Circuit Breakers, Motor Controllers, and Fuses
120	Control Transformers and Resistors
125	Motors and Generators
128	Ventilating Piping
130	Luminaires
135	Utilization Equipment
140	Flexible Cords
145	Receptacles and Attachment Plugs
150	Signaling, Alarm, Remote Control, and Communications Systems

Figure 5-2. *Parallel numbering can be used to help find or compare the requirements in the relevant articles.*

Class III, Division 1
- Locations in which easily ignitable fibers or materials producing combustible flyings are handled, manufactured, or used

Class III, Division 2
- Locations in which easily ignitable fibers are stored or handled other than in the process of manufacture

Articles 501 to 503
Articles 501, **502**, and **503** cover Class I, II, and III hazardous locations in further detail. These articles have their own parallel numbering system, as was the case in **Chapter 3** of the *NEC* with raceways. **See Figure 5-2.** Not all section topics will be identical from **Article 501** to **502** and **503**, as some hazardous locations have hazards unique to their class. However, where the parallel numbering system does apply, it aids the user in finding the relevant section quickly.

Article 504
Article 504 covers intrinsically safe systems. Although it is located after **Articles 501** to **503**, which use the Class and Division method of classifying hazardous locations, and before **Articles 505** and **506**, which use the alternative Zone method of classification, **Article 504** applies to all groups and both methodologies.

An intrinsically safe (IS) circuit is a circuit that is not capable of producing a spark or thermal effect that will cause ignition under specified test conditions. The use of IS circuits and systems can reduce the need for explosion-proof equipment and wiring methods. Note that the control drawing must be followed according to **504.10(A)**, and the IS conductors must be kept separate from other non-IS conductors. Bonding is also of critical importance. See **504.60**.

Articles 505 and 506
Article 505, Zone 0, 1, and 2 Locations, has similar requirements as **Article 501** and relates to installations in locations containing flammable gases

or vapors. **Article 506, Zone 20, 21, and 22 Locations for Combustible Dusts or Ignitible Fibers/Flyings**, has similar requirements as **Articles 502** and **503** and contains information for the installation of wiring in locations where combustible dusts or ignitable fibers or flyings may be present.

In **Articles 501** to **503**, the category *division* is used to classify locations based upon the likelihood of a hazard being present; however, this method provides only two categories. The zone method used in **Articles 505** and **506** adds a third category. This third category is the *hazardous condition* and is present continuously. Zone 0 and 20 are examples of this condition. An example of a Zone 0 location would be inside a vented fuel tank.

Specific Hazardous Locations

Article 510, with only two paragraphs, is the shortest article in the *NEC*. The scope of **Article 510** introduces **Articles 511** through **517** and shows how the general rules of **Articles 501** to **504** apply to the occupancies within **Articles 511** through **517**. Note that these subsequent articles may modify requirements in **Articles 501** to **504**.

Each of the articles in **511** through **517** covers particular hazardous occupancies, such as aircraft hangers, gas stations, and health care facilities.

Articles 511 to 516

The occupancies in **Articles 511** to **516** bring forth their unique issues that are hazardous with the misuse or incorrect installation of electrical equipment. **Article 511** offers guidance to what areas within a repair garage warrant classification using **Articles 500** through **506**.

Article 513, Aircraft Hangars, covers locations used to fuel, repair, and paint aircraft. Areas around the aircraft or fuel tanks are classified locations. This article contains requirements unique to these locations.

Article 514, Motor Fuel Dispensing Facilities, commonly called "gas stations," shows what areas are classified in these occupancies. A concern for these areas stems from the fact that they are readily accessible to the public. Disconnecting means at unattended facilities must be readily accessible for the use of the customers. See **514.11(C)** in the *NEC*.

Electrical installations in larger facilities used for the transfer or mixing of flammable liquids from pipelines and tank trucks are covered in **Article 515**. In a typical gas station, standardized gas pumps are used to dispense fuel. **Article 515** contains requirements and guidance where a multitude of methods can be used to move hazardous liquids between containers and pipelines.

Combustible liquids or powders are used in spray and dipping operations in paint booths and printing processes. Lighting, heating for drying areas, and other wiring in these locations must be installed according to **Article 516**.

Article 517, Health Care Facilities

Health care facilities are those that occupy buildings or portions of buildings in which medical, dental, psychiatric, nursing, obstetrical, or surgical care is provided. The definition of *health care facilities* in **Section 517.2** outlines locations covered by this article. These locations include, but are not limited to,

The electrical requirements of a hospital involve complex systems of power, communications, and life safety.

hospitals, nursing homes, limited care facilities, clinics, medical and dental offices, and ambulatory care centers, whether permanent or movable.

Fact

NFPA 99: Health Care Facilities Code provides many specific design, installation, and maintenance rules required for health care facilities.

Article 517 is included under the scope of "specific hazardous locations" in **Article 510** because there are requirements for flammable anesthetics within **Article 517**. This dates back to the 1953 edition of the *NEC*, in which a section within **Article 510** titled *Combustible Anesthetics* appeared. At that time, flammable anesthetics such as cyclopropane were widely used, and accidents were relatively common. In 1959, during a restructuring of the *NEC*, Article 517, Flammable Anesthetics, first appeared. Several editions later in 1971, **Article 517** became **Health Care Facilities**. The scope then expanded to cover the entire facility, not just the anesthetizing locations. Today in the United States, flammable anesthetic gases are not allowed for use in health care facilities. However, some other countries may still allow their use, precluding the removal of the section entirely.

Figure 5-3 Informational Note Figure 517.31(a)

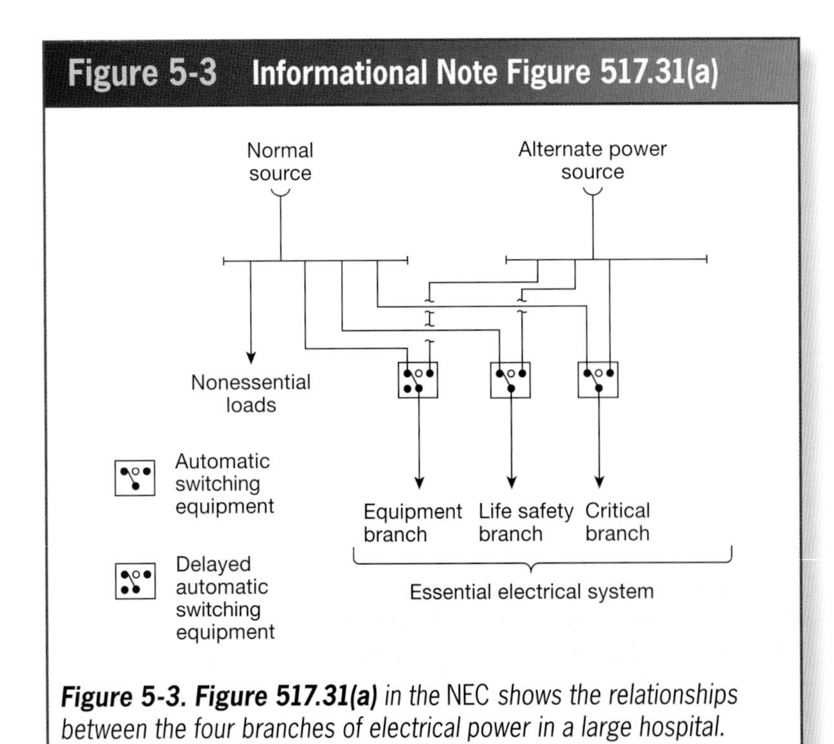

***Figure 5-3. Figure 517.31(a)** in the* NEC *shows the relationships between the four branches of electrical power in a large hospital.*

Fact

Article 517 applies only to those health care facilities that provide services to people; it does not include veterinary facilities.

One of the most important sections in **Health Care Facilities** is **517.2, Definitions**. Some key definitions that directly impact the work of an Electrical Worker can be found in this section.

Critical Branch, Equipment Branch, and Life Safety Branch

The electrical system of a hospital is divided into two parts: the essential electrical system and the nonessential loads. The essential electrical system is further broken down into three branches: the critical branch, the life safety branch, and the equipment branch. **See Figure 5-3.**

All of the loads that comprise the essential electrical system are required to have an automatically switched alternate source of power. These sources of power are usually provided by generators, but as of the 2017 *NEC*, fuel cells are now included due to their proven reliability. See **517.33** and **517.34** in the *NEC* to determine what kinds of loads are specifically allowed on the life safety and critical branches.

Patient Care Space

The persons who have the legal responsibility for the operation of the facility, known as the *governing body*, decide what kind of patient care takes place within each area within the facility and designate the spaces accordingly. There are four kinds of patient care spaces, each of which has unique wiring and

Category 1 (Critical Care) Space. Space in which failure of equipment or a system is likely to cause major injury or death of patients, staff, or visitors. [99:3.3.136.1]

Informational Note: Category 1 spaces, formerly known as critical care rooms, are typically where patients are intended to be subjected to invasive procedures and connected to line-operated, patient care–related appliances. Examples include, but are not limited to, special care patient rooms used for critical care, intensive care, and special care treatment rooms such as angiography laboratories, cardiac catheterization laboratories, delivery rooms, operating rooms, post-anesthesia care units, trauma rooms, and other similar rooms. [99:A.3.3.136.1]

Category 2 (General Care) Space. Space in which failure of equipment or a system is likely to cause minor injury to patients, staff, or visitors. [99:3.3.136.2]

Informational Note: Category 2 spaces were formerly known as general care rooms. Examples include, but are not limited to, inpatient bedrooms, dialysis rooms, in vitro fertilization rooms, procedural rooms, and similar rooms. [99:A.3.3.136.2]

Category 3 (Basic Care) Space. Space in which failure of equipment or a system is not likely to cause injury to the patients, staff, or visitors but can cause patient discomfort. [99:3.3.136.3]

Informational Note: Category 3 spaces, formerly known as basic care rooms, are typically where basic medical or dental care, treatment, or examinations are performed. Examples include, but are not limited to, examination or treatment rooms in clinics, medical and dental offices, nursing homes, and limited care facilities. [99:A.3.3.136.3]

Category 4 (Support) Space. Space in which failure of equipment or a system is not likely to have a physical impact on patient care. [99:3.3.136.4]

Informational Note: Category 4 spaces were formerly known as support rooms. Examples of support spaces include, but are not limited to, anesthesia work rooms, sterile supply, laboratories, morgues, waiting rooms, utility rooms, and lounges. [99:A.3.3.136.4]

receptacle requirements.

Notice that many definitions in **Article 517** include a reference to *NFPA 99: Health Care Facilities Code*, the governing document for health care facilities. Also take note of the terms in parenthesis after the category designation; these are older terms that are being phased out and replaced with category terminology. By the 2023 edition of the *NEC*, the *NEC* will only include the terms that align with the terminology used in *NFPA 99*. This gradual process allows users to adapt to the new terminology over several *Code* cycles, making the *NEC* a more user-friendly document.

Patient Care Vicinity

The definition of *patient care vicinity* has measurements associated with it, and as a result it influences how wiring is done in a patient care space. Patient care vicinities include an area six feet from the patient's bed, chair, or any other device, such as a treadmill, that the patient may use during an examination. This area extends to a height of seven feet and six inches above the floor. See **517.2** for the complete definition.

Wiring Methods

The wiring methods in a health care facility must comply with **Chapters 1** through **4** and **Article 517**. **Section 517.13** has further wiring requirements for patient care spaces. The wiring must be located in a raceway or part of a cable assembly with a sheath or armor that is capable of performing as an equipment grounding conductor per **250.118**. Standard metal-clad (MC) cable does not have armor that is capable of being used as an equipment grounding conductor. Specially listed MC cable, as stated in **250.118(10)(b)** or **(c)**, is usually referred to as "hospital grade MC" by tradespeople.

The other requirements in **250.118(10)** apply as well. These raceways and cables must have an insulated equipment grounding conductor installed along with the branch-circuit conductors supplying the patient care space. This additional grounding serves to reduce voltage potentials in the area of the patient, increasing electrical safety.

Fact

The additional grounding conductor required in **Section 517.12** is often referred to as a "redundant ground."

Sections 517.16 through **517.20** contain receptacle, branch-circuit, and ground-fault protection requirements. Isolated ground receptacles can have a different potential to ground than the immediate area, and are therefore not allowed in patient care vicinities. They may be installed elsewhere within the facility, however, such as at a nurse's station.

The multiple receptacles required in Category 2 and Category 1 spaces may appear extraneous, but these receptacles provide the flexibility needed to supply whatever equipment is needed for patient care and allow the equipment to be arranged within the space accordingly. For example, an operating room requires a minimum of 36 receptacles divided between at least two branch circuits.

A special grade of receptacle is required in some Category 1 and 2 spaces as well as in unclassified anesthetizing locations. A hospital-grade receptacle is tested and listed differently than a commercial-grade receptacle. Among other things, a hospital-grade receptacle has a more robust grounding connection. **See Figure 5-4.**

Section 517.160 outlines the requirements for isolated power systems (IPS), a feature unique to hospitals. An IPS is an isolated, ungrounded system. In the early days, IPSs were used to handle issues associated with flammable anesthetics. Since the system is ungrounded, arcing from a ground fault is lessened. This in turn reduces the probability of ignition of a flammable anesthetic.

However, the primary reason for the use of IPSs today is the continuity of power provided by these systems. Continuity of power is maintained to supply life support equipment and other life-critical components necessary for the care of the patient. While the IPS is ungrounded, the current is monitored for leakage. This monitoring device operates in the range of five milliamperes. When leakage current beyond what is acceptable is detected, the IPS annunciates the status, and personnel can take appropriate action to correct the situation.

It is very important to pay attention to the installation requirements of an

Figure 5-4 Receptacle in a Patient Care Location

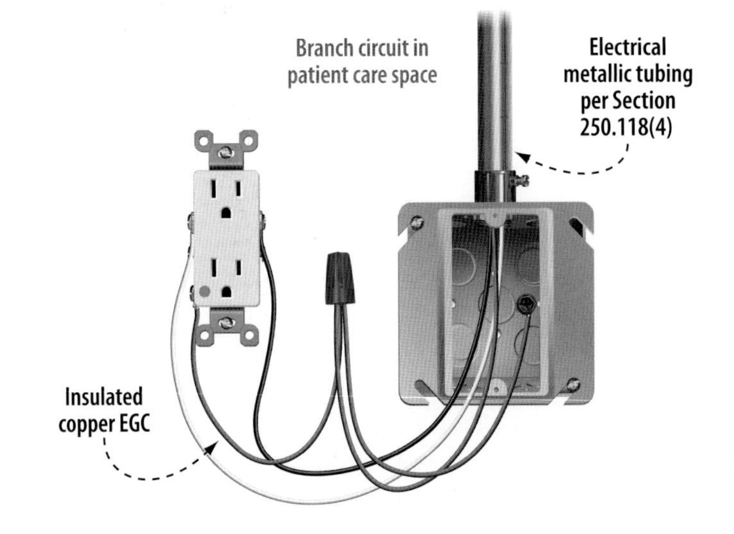

Figure 5-4. *Receptacles in patient care locations have special grounding requirements.*

IPS. The wiring of an IPS supplying the devices in the area is color-coded orange, brown, and yellow for line voltages, and each conductor is further identified with a colored stripe. These systems require very high dielectric values for the insulation on the conductors. For this reason, Type XHHW is typically specified.

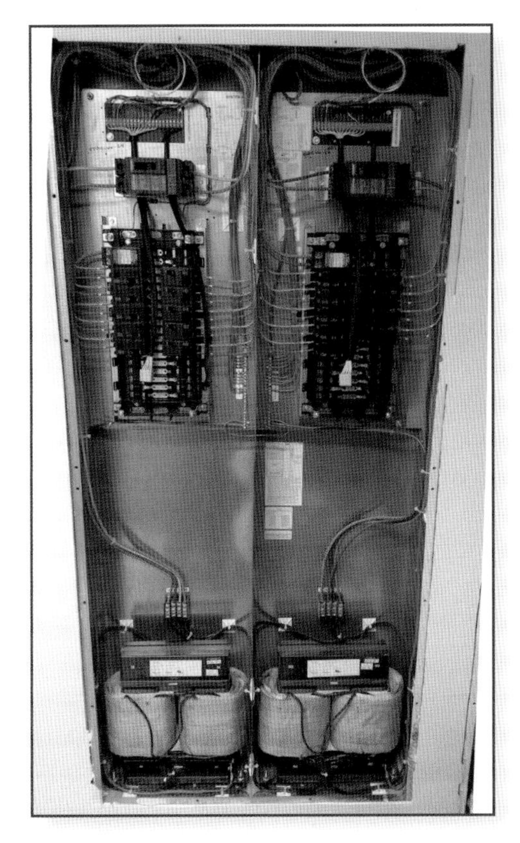

IPS panels may be used to power an operating room. Note the bonding conductor between the panels. Photo courtesy of Brian McDonald

The electrical infrastructure of a modern hospital has features similar to those in commercial and light industrial applications. The requirements for back-up power, emergency systems, gas systems, communication systems, data systems, fire alarm systems, and other systems offer a challenge to even experienced Electrical Workers.

Assembly Areas

As explained in **Section 518.1**, the scope of **Article 518** covers buildings or portions of buildings designed to host gatherings of 100 or more people for purposes such as worship, eating and drinking, and amusement.

Meeting areas are classified in the NEC *as places of assembly.*

Examples of assembly areas, listed in **518.2**, include such areas as churches, synagogues, mosques, conference rooms, and gymnasiums. See **518.2** for a more exhaustive list. Note that theaters, though they are designed to hold large numbers of people, are covered separately in **Article 520**.

The over-arching concern for places of assembly is the need to be able to evacuate many people in the event of a fire. For this reason, the wiring methods in **Article 518** exclude nonmetallic raceways unless encased in concrete. Nonmetallic wiring methods can burn and produce large amounts of smoke, which make the space untenable and reduce the chances of safe evacuation by obscuring vision and exposing occupants to large quantities of toxic smoke.

Entertainment Venues

Entertainment venues such as theaters and carnivals have special requirements. Carnivals are largely outdoor events where protection of the public is paramount. Theaters have similar, though not identical, concerns as places of assembly, including the need to address permanent and temporary wiring for events.

Article 520, Theaters

Theaters present particular concerns that are not found in other occupancies regarding the safety of the audience, performers, and crew involved in the presentation of live plays and performances. Permanent and mobile switchboards, such as those that a musical group may bring with them for a performance, are addressed in this article, along with other permanent and portable stage equipment. Much of the equipment involved in plays and other performances requires frequent interchange, which necessitates the use of flexible cords and the guarding of equipment.

Articles 522 and 525, Amusement Attractions

Control systems for permanent amusement attractions are addressed in **Article 522**, and requirements for the wiring of carnivals, circuses, and fairs are covered by **Article 525**.

It is important to stress that **Article 522** has a very focused scope; it is only concerned with the control systems of permanent amusement attractions. These control circuits are limited to voltages 150 volts AC or 300 volts DC to ground. Note once again this is only for the control circuits, and does not include power to the motors and other equipment that may be controlled by the circuits. Per **Section 522.7**, only qualified personnel shall be allowed to work on these rides and attractions.

Article 522 first appeared in the 2008 *NEC* to provide consistency and a standard for inspectors concerning the specialized wiring practices used in large amusement parks. In order to manage their rides and to ensure the safety of their guests, large amusement parks use complex control systems that rival those found in large industrial applications.

Article 525 is primarily concerned with temporary or transient events such as a traveling carnival setting up its rides, booths, and other attractions. This article refers to **Articles 518** and **520** for the requirements for any permanent structures that may be a part of an amusement venue.

Part III, Wiring Methods, contains requirements for cords and cables. For example, splices are not allowed between boxes or fittings, and all rides must have a disconnect within six feet of the operator's location when running the ride. GFCI protection is required wherever the general public has access, with the exception of egress lighting, which shall not be GFCI protected. Note that since **Article 525** resides in **Chapter 5**, **Section 90.3** requires the GFCI requirements found in **Chapter 2** to be applied unless modified elsewhere in **Chapter 5**.

Article 530, Motion Picture and TV Studios

Many of the topics of concern in **Article 530** are the same as or similar to those in **Article 520**, which covered theaters. **Article 530** applies to all studios where television shows or movies are made, whether they use digital or film media. **Article 530** also includes requirements for film storage.

Article 540, Motion Picture Projection Rooms

Celluloid film is extremely combustible, and when it catches fire, it is nearly impossible to extinguish. For this reason, separate professional projection rooms became standard, and **Article 540** was added to the *NEC*. Projection rooms are required to be constructed with a shutter that could be closed over the glass windows, separating the audience from the fire. A quick glance at **Part II** for professional projectors and **Part III** for non-professional projectors reveals many more requirements for those projectors that require a projection room. However, the manufacture of celluloid films largely stopped in 1951 and has since been replaced by the far less combustible substitute acetate and, in more modern times, digital media.

Agricultural Buildings

Article 547 contains the requirements for electrical systems in buildings that are used for agricultural purposes, including feeding animals. Even very low voltages and currents can alter milk production in dairy cows and reduce water and feed intake in other animals, affecting their health. Some of these buildings, such as dairies, are frequently washed down, qualifying them as damp or wet areas. In addition, these buildings can have notable amounts of dust related to the feeding and care of the animals.

A major part of **Article 547** is the establishment of equipotential planes. Equipotential planes are utilized in agricultural areas to minimize voltage potentials, as animals are very sensitive to voltages, and milking machines and other electrical equipment such as automated feeders can expose an animal to harm. The equipotential plane is an area where wire mesh or other conductive elements are placed under or within concrete and bonded to all the metal structures and equipment that may become energized, such as stanchions, stalls, and electrical equipment. See **547.2** in the *NEC*.

In addition to the protection of personnel, the requirements for GFCI-protected receptacles found in **547.5(G)** may also benefit the livestock.

Section 547.5 also covers the additional requirements beyond **Chapters 1** through **4** of the *NEC* for agricultural buildings, addressing the concerns of dust, corrosive conditions, and wet or damp locations. Note the additional requirements for cable support as well.

Manufactured Buildings, Dwellings, and Recreational Vehicles

These occupancies are manufactured offsite and as such are located in one area of the *NEC*. The *NEC* contains requirements for the manufacturers of these products and requirements for installers connecting to or providing power for these units.

Articles 545 and 550

In the 2020 edition of the *NEC*, **Article 550** was reorganized into two articles:

- **Article 545, Manufactured Buildings and Relocatable Structures**
- **Article 550, Mobile Homes, Manufactured Homes, and Mobile Home Parks**

Requirements previously included in **Article 550** regarding non-dwelling unit structures have been relocated to a new article, **Article 545**, in the 2020 edition. This was done to clarify the requirements for manufactured buildings that are not used as dwellings. This clarification enables the *NEC* to properly address structures such as mobile offices, restrooms, classrooms, break rooms, and other relocatable structures not used as dwellings. Much of the information in the new article was taken from the section of **Article 550** titled "Mobile Homes Used as Other Than Dwelling Units" in earlier editions of the *Code*.

Typical manufactured homes are covered under ***Article 550****.*

Article 550 now contains only the requirements for manufactured homes and mobile home parks. In the context of **Article 550**, manufactured and mobile homes are treated the same. The requirements for the manufacture of these homes include receptacle and branch circuit requirements that are similar, but not identical, to those found in **Article 210, Branch Circuits**. Another requirement that needs to be fulfilled is the mandated dielectric testing of a manufactured home; this and other tests are covered in **550.17**.

Mobile home parks are covered in **Part III** of the article. **Section 550.31** shows the allowable demand factors for a park, with each lot having a minimum of 16,000 volt-amperes allocated for the future home to be placed there. These demand factors represent the load diversity present in a park, as it is unlikely for every tenant to draw large currents

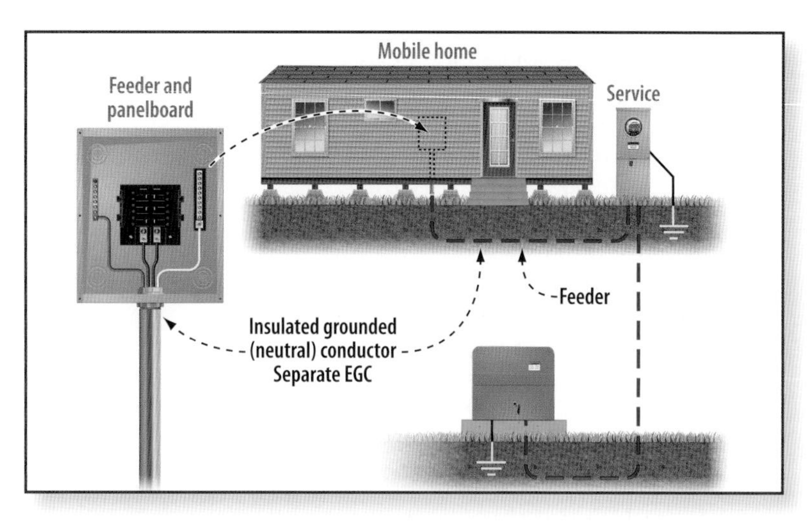

Services for manufactured homes are usually located outside the home, making the power supply to the home a feeder.

associated with daily living at the same time. Such demand tables are common throughout the *NEC*, where load diversity can be taken advantage of to minimize feeder or service requirements.

Article 551, Recreational Vehicles and Recreational Vehicle Parks

NEC **90.2(B)** clearly states that the *NEC* does not cover automobiles, with the exception of recreational vehicles (RVs). A recreational vehicle may be a trailer type or one that can be driven. **Article 551** covers the equipment and wiring in a recreational vehicle other than the circuits that are involved in the automotive vehicle portion of the RV. This article also contains the information necessary for connection of recreational vehicles to electrical supply and the installation of equipment related to electrical installations in RV parks.

Note that with the exception of "Recreational Vehicle," the definitions in **551.2**, while sounding very similar, if not identical, to other definitions in the *NEC*, apply only to **Article 551**.

The wiring that a manufacturer of recreational vehicles must provide is covered in **Part IV** of the article. This includes branch circuits, receptacle requirements, switches, luminaires, power supplies, and panelboards as well as the means with which the RV is connected to the supply. **Section 551.60** details the high-voltage testing an RV must undergo, similar to a manufactured home.

Part VI covers the requirements and allowances for RV parks. **Section 551.71** contains the receptacle requirements for an RV park, which stipulate what percentage of the sites must have a certain amperage of receptacles available. Notice that where an RV site has a 50-ampere receptacle, it must also have a 30-ampere receptacle. This was added in the 2017 edition of the *NEC* to prevent a person with an RV that requires only 30 amperes from using an adapter to feed his or her RV with a 50-ampere source.

Section 551.73 categorizes the demand for the receptacles by type, and **551.73(B)** shows the demand factors that may be used when more than one site is connected to a feeder or service.

Article 552, Park Trailers

A *park trailer* by definition is a trailer with an area of 400 square feet or less once it is set up. Park trailers are intended for seasonal use, and are not intended to be used as permanent dwellings, offices, or for any other purpose on a permanent basis. Much of **Article 552** is similar to **Articles 550** and **551**.

Structures Near Bodies of Water

Electrical shock drownings have been documented around marinas and boatyards. In these tragedies, electrical current is usually present in the water itself, and people swimming in these areas may lose muscle control or experience increased levels of fatigue and drown as a result. Proper wiring around marinas, boatyards, and other structures near bodies of water can reduce or eliminate the hazards of electrical shock drownings and prevent these tragedies.

The scope of **Article 555**, which covers marinas, boatyards, floating buildings, and commercial and noncommercial docking facilities, includes wiring of equipment in areas comprised of fixed or floating piers, wharves, docks, floating buildings, and other areas in marinas, boatyards, boat basins, boathouses, yacht clubs, boat condominiums, and docking facilities associated with single-family dwellings, two-family dwellings, multifamily dwellings, and residential condominiums as well as any multiple docking facility or similar occupancy. The scope also includes facilities intended for the use or purpose of the repair, berthing, launching, storage, or fueling of small craft and the moorage of floating buildings.

The *electrical datum plane* is a definition that must be understood in order to properly apply **Article 555**. Water levels rise and fall, depending on the body of water in question, according to factors such as seasonal changes, tidal fluctuations, and the operation of hydroelectric power plants. The electrical datum plane for a given body of water provides a reference relative to the maximum water level likely to be encountered for the body of water. All receptacles, transformers, and other equipment must be located above the electrical datum plane. Since this definition is used in two articles, **Articles 555** and **682**, it resides in **Article 100**. The actual distances for equipment from the electrical datum plane are article-specific. See **555.3** for measurements for mounting equipment within the scope of this article.

As would be expected near a body of water, ground-fault protection is required in **Article 555**. There are two kinds of ground fault protection used around these locations. **Section 555.33** has the requirements for providing shore power and for uses other than shore power. Receptacles that provide shore power are required to have ground-fault protection not exceeding 30 milliamperes. This is significantly higher than the GFCI protection required for receptacles that are used for other than shore power. See **555.35(A)(1)** in the *NEC*. Note that this 30-milliampere ground-fault protection requirement was introduced in the 2017 edition of the *NEC* and is a reduction from earlier editions, designed to reduce the electrical shock drowning hazard. The allowed milliampere value is higher for feeders and branch-circuit conductors. See **555.35(A)(3)**.

Per **555.10**, permanent safety signage must be installed to notify people of the shock hazards associated with swimming near a marina, docking facility, or boatyard. These signs must be posted at all approaches to the area of concern, including approaches from both the land and water.

Temporary Installations

"It's temporary" is a frequent excuse unqualified installers may use to improperly lower wiring standards and requirements for any installation that is not permanent. In reality, while temporary installations have different standards, some of which are not as strict, other requirements for temporary installations are identical to or even more restrictive than requirements found elsewhere in the *Code*.

NEC **590.2(A)** states:

> Except as specifically modified in this article, all other requirements of this *Code* for permanent wiring shall apply to temporary wiring installations.

The rules for ampacity of conductors and overcurrent protection still apply. Receptacles on circuits used on construction sites may not share the circuit with temporary lighting. This way, if an overcurrent protective device operates due to a power tool overloading the circuit, the job site is not thrown into darkness. See **590.4(D)(1)** in the *NEC*.

Section 590.6 requires either GFCI protection or an assured equipment grounding conductor program that is continuously enforced by designated persons. The interval for such testing of the equipment grounding conductor is to be no more than three months.

Temporary locations are treated as their own occupancy type from a *Code* perspective. Wherever temporary wiring for construction or holiday purposes is installed, compliance with **Article 590** is necessary. Temporary wiring for construction must be removed upon the completion of construction, while holiday lighting must be removed after 90 days. Nonmetallic sheathed cable may be used to provide temporary wiring without the restrictions found in **334.10** and **334.12**. Open splices are allowed if the rules found in **590.4(G)** are followed.

SUMMARY

The first chapter in the special chapters of the *NEC* is **Chapter 5**. Whenever an Electrical Worker is installing or maintaining electrical equipment in any of these areas, the appropriate article must be consulted. Each article has its own requirements for the protection of people and property so as to safeguard against the hazards unique to the occupancy. **Chapter 5** is organized generally by occupancy type and the hazards involved. This organization aides the user in finding the information that is relevant to the task at hand. In addition to the requirements in **Chapter 5**, the other chapters of the *NEC* apply to every installation unless they are modified by the requirements found in **Special Occupancies**.

QUIZ

Answer the questions below using a copy of the *NEC* and by reviewing Chapter 5 of this text.

1. **What article covers electrical installations specific to hospitals?**
   a. **Article 501**
   b. **Article 513**
   c. **Article 517**
   d. **Article 518**
2. **How far away from a gas pump at a gas station can the area be a classified location?**
   a. 18"
   b. 10'
   c. 20'
   d. The area outside the dispenser is not classified.
3. **A wall or arch that separates the stage from the auditorium where the audience is seated is known as the _?_.**
   a. dais
   b. proscenium
   c. stage boundary
   d. stage set
4. **Twenty RV sites that are equipped with 50-ampere receptacles have a total demand of _?_.**
   a. 1,000 VA
   b. 32,400 VA
   c. 108,000 VA
   d. 240,000 VA
5. **Decorative lighting for holidays is permitted to be installed for a period of time not to exceed _?_.**
   a. 30 days
   b. 30 days past the date of the holiday
   c. 60 days
   d. 90 days
6. **Other NFPA documents that relate to installations in aircraft hangers include _?_.**
   a. *NFPA 13*
   b. *NFPA 20*
   c. *NFPA 409*
   d. *NFPA 720*
7. **Panelboards that serve the same patient care vicinity must be bonded together with a _?_.**
   a. #10 AWG or larger bare conductor
   b. #10 AWG or larger insulated conductor
   c. raceway complying with 250.118
   d. any one of the above
8. **Critical care areas are also known as _?_.**
   a. Category 1 spaces
   b. Category 2 spaces
   c. Class 1 locations
   d. Class 2 locations
9. **In a gymnasium that seats 1,000 people, which of the following wiring methods are not allowed?**
   a. EMT conduit
   b. ENT conduit encased in 2 inches of concrete
   c. PVC conduit
   d. MI cable
10. **The ampacity of a #24 AWG copper control circuit conductor for a permanently installed roller coaster is _?_.**
    a. 0.5 A
    b. 0.8 A
    c. 1.19 A
    d. 2 A

CHAPTER 6: SPECIAL EQUIPMENT OF THE *NEC*

Some equipment has additional requirements, or in some cases less stringent requirements, that cannot be addressed generally in the other chapters of the *NEC*. Such equipment may have short-term demands of high current, special bonding requirements, unique voltage supplies, or additional GFCI protection. Electrical Workers who routinely install particular kinds of equipment may become familiar with and even memorize the requirements for that equipment, but it is important for all Electrical Workers to be aware of the topics within **Chapter 6, Special Equipment**, so when they encounter special types of equipment they will know how to access the information they need.

Objectives

» Recall the titles of the articles found in **Chapter 6** of the *NEC*.

» Locate definitions applicable to **Chapter 6** of the *NEC*.

» Apply articles, parts, and tables to identify specific *NEC* **Chapter 6** requirements.

Chapter 6

Table of Contents

CHAPTER 6 OF THE *NEC*

Many, though not all, of the articles in **Chapter 6** are organized by kind or topic. **Articles 600** to **605** are focused on equipment found in most commercial occupancies. **Articles 610** and **620**, which discuss cranes and elevators respectively, are concerned with equipment that uses motors. **Articles 625** and **626** both relate to transportation, covering electric vehicles and requirements for supplying power to truck parking spaces for trucks with refrigerated trailers. Electronic equipment such as audio, information technology (IT), and other sensitive electronic equipment is covered by **Articles 640**, **645**, **646**, and **647**. Industrial equipment is addressed within **Articles 660** through **670**. Requirements for installing equipment where exposure to water hazards is involved are located in **Articles 680** and **682**, and integrated electrical systems are covered in **Article 685**. Alternative sources of power such as solar, wind, and fuel cells are covered as well. **Article 695, Fire Pumps**, closes out the chapter with its own area of **Chapter 6**.

ARTICLE 600, ELECTRIC SIGNS AND OUTLINE LIGHTING

Electric signs and outline lighting are common to many commercial and industrial buildings. Where a commercial occupancy is accessible to pedestrians, **Section 600.5** requires an outlet to be installed for a sign not intended to serve employees or people who may be servicing the place of business. The circuit is required to be at least 20 amperes and serve no other load.

There are two parts to **Article 600**. In **Part I**, general requirements such as disconnects, circuits, grounding and bonding, and issues involving the location of the sign and working spaces are covered. **Part II** contains the requirements for skeleton tubing and outline lighting, and also applies to the circuits on the secondary side of neon transformers. *Skeleton tubing* is the neon tubing that can be bent to form letters, numbers, and symbols for a neon sign. Neon signs can operate at 5,000 volts or more.

The grounding and bonding requirements for electric signs are unique. *Bonding* is connecting conductive parts so they are electrically continuous. This bonding is required to reduce the capacitive coupling that can occur with higher voltages and frequencies. Secondary circuit length is limited to reduce this capacitive effect. See **600.32(J)**. Capacitive coupling can reduce the brightness of the neon sign and overload the power supply. When non-metallic raceways are used, the bonding conductor is installed separately and spaced away from the raceway. This spacing minimizes insulation stress on the secondary conductors. The high-voltage secondary conductors are typically gas, tube, and oil (GTO) cable as mentioned in **600.32(B)**.

In **600.7(B)(2)**, flexible conduit is allowed as a bonding means in lengths up to 100 feet. Bonding conductors used to join enclosures must be at least size 14 AWG. When used to route the conductors from the secondary side of a neon transformer, the flexible conduit is not to be used as a fault return path as it would be in small branch-circuit applications where the length would be limited to six feet. Although the secondary of the transformer may operate at 5,000 volts or more, the current would be in the milliampere range.

ARTICLES 604 AND 605, MANUFACTURED WIRING SYSTEMS AND OFFICE FURNISHINGS

A *manufactured wiring system* is a system of components that is used to connect equipment with prewired polarized connectors, most commonly cable assemblies. These assemblies are ordered from the manufacturer in the lengths required for the planned installation.

Manufactured wiring systems are used to connect luminaires in accessible ceilings and receptacles in raised floors and, where made accessible, within walls. **Section 604.10** permits these systems in accessible dry locations. Manufactured wiring systems may also be comprised of flexible cords, busways, and raceways. See **604.100**.

Office furnishings such as cubicles are commonly fed with power and communication circuits. Office furnishings may be cord-connected, as outlined in **605.5**, or they may be permanently connected to the building electrical system using one of the applicable methods in **Chapter 3**.

CRANES, ELEVATORS, AND ESCALATORS

Cranes, elevators, and escalators are used to move objects or people. All of these types of equipment use motors to get the job done; therefore, **Articles 610** and **620** often refer to **Article 430, Motors.**

Cranes have concerns unique to their use that involve not only carrying a load, but also holding it in place. The safety concerns of elevators and escalators are obvious whenever the transportation of people is involved. Most jurisdictions mandate routine inspections of elevators, moving walks, and escalators.

Article 610, Cranes and Hoists

Article 610, Cranes and Hoists, has much in common with **Article 430, Motors, Motor Circuits, and Controllers**, and also has requirements that are necessary and particular to cranes. When a motor is started, there is high current. For a motor that is used continuously, this brief high inrush current is not a factor in the heating of the conductor. A crane motor, however, is not used continuously, and can have multiple stop-and-start operations in a short period of time. This produces higher current on average than a motor that is used continuously.

In **430.22(E)**, branch circuits for motors in different duty applications

Cranes and hoists of all sizes are covered in ***Article 610****.*

are sized using the motor nameplate. The same logic is used when sizing motor conductors for cranes. Further demand factors may be applied for multiple motor applications. See **610.14(E)(1)** and **(2)**.

This article also contains ampacity tables that are to be used in place of the ampacity tables found in **Chapter 3**. The ampacities in **Table 610.14(A)** are used due to the short time that conductors are energized when supplying crane motors.

Article 620, Elevators

For tall buildings to be practical, elevators are essential. **Article 620** covers elevators, dumbwaiters, escalators, moving walks, platform lifts, and stairway chairlifts.

Fact

In the 1850s, Elisha Otis invented the safety elevator. If the cable that supported the elevator car broke, the car would not fall. A few decades later the first electric elevator was created.

In addition to the requirements of the *NEC*, elevators are subject to the mandates developed by the American Society of Mechanical Engineers (ASME) in the standard *ASME 17.1*, while stairway lifts and platform lifts are subject to the additional requirements in *ASME 18.1*.

Fact

Chapter 21 of *NFPA 72: National Fire Alarm and Signaling Code* also contains information of which Electrical Workers should be aware regarding elevator control.

ASME 17.1 deals mostly with the construction and maintenance of elevators, but it has a few requirements of interest to the Electrical Worker, such as machine room lighting requirements and methods of interfacing elevator controls with the fire alarm system. The interface with the fire alarm system is used to recall the elevator to predetermined floors if smoke is detected in the lobbies, machine room, or hoistway. Heat detectors are also used in the hoistway and in the machine room to disconnect power before sprinklers activate in these areas, as water and electricity do not mix well, so that the elevator braking mechanism is not compromised. If coordinated properly, the elevator's braking mechanisms can safely secure the car before sprinkler water is introduced into the system.

Figure **620.13** is useful in navigating this article. Notice that **Articles 430** and **450** are used for elevator applications. **See Figure 6-1.**

Tip

Writing the page numbers adjacent to each of the *Code* references shown in **Informational Note Figure 620.13,** creating a table of contents just for this article, can be helpful. Check with the local testing authority to verify that such notes are allowed before attempting to use the same copy for a licensing or certification exam.

Figure 6-1 Informational Note Figure 620.13

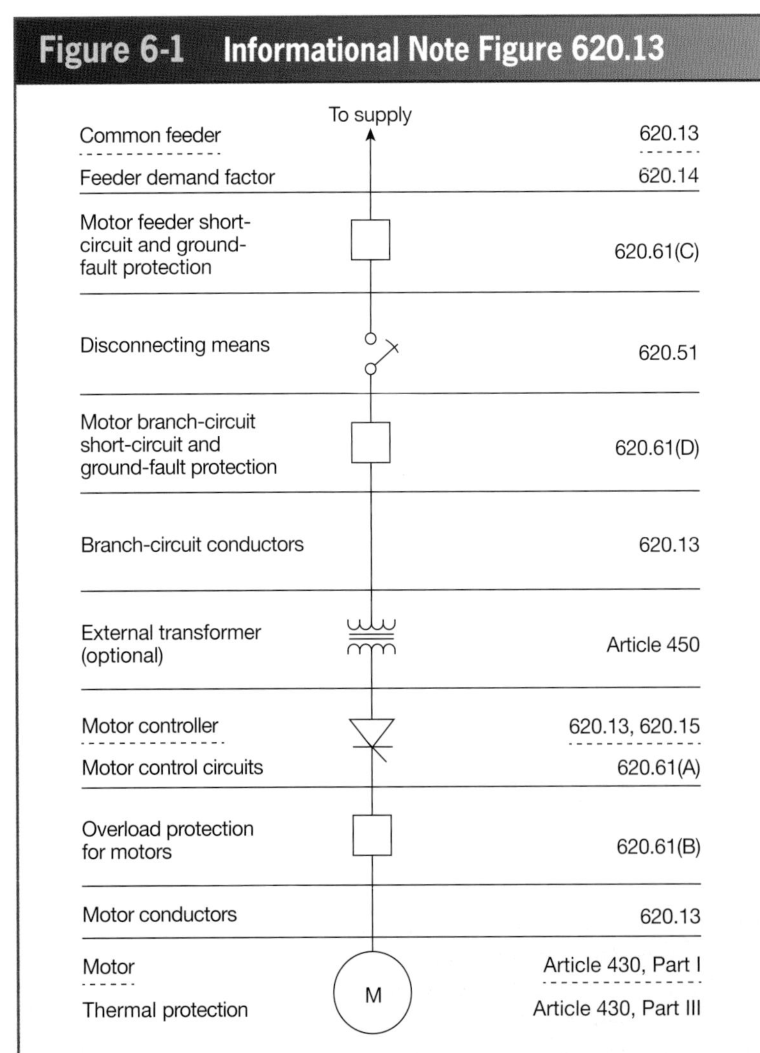

Figure 6-1. *A one-line diagram is included in the NEC to aid the user of* ***Article 620*** *in finding the relevant section.*

Part III of **Article 620** covers wiring. Certain equipment and devices need to be on separate branch circuits in order to facilitate maintenance of the elevator. The following devices and equipment need to be on separate branch circuits:

- **620.22**: lighting, receptacles, and ventilation for each elevator car
- **620.23**: lighting for machine rooms and control spaces
- **620.23**: receptacles for machine rooms
- **620.24**: receptacles and lighting in hoist ways and elevator pits

Furthermore, **620.37** in **Part IV** only allows the wiring that is directly used for the elevator, dumbwaiter, and other elevator-related items to be located in the hoistway, machine room, or control room spaces. See **620.37**.

ARTICLE 625, ELECTRIC VEHICLE POWER TRANSFER SYSTEM

Section 625.1 explains that the scope of the article covers electric vehicles and does not include forklifts and other industrial trucks. The informational note to this section offers guidance by referring the reader to *NFPA 505: Fire Safety Standard for Powered Industrial Trucks Including Type Designations, Areas of Use, Conversions, Maintenance, and Operations* for additional information.

The definition of *electric vehicle* is found in **Article 100** and describes the kind of vehicle and equipment that would be covered in the article. The definitions within **625.2** are critical to the understanding of how to use this section of the *NEC*.

Concerns about the manner in which electric vehicles are charged were first addressed in the 1996 edition of the *NEC*. At that time, the batteries used in electric vehicles were mostly the lead acid type, which could give off hydrogen and other gasses. This led to safety concerns regarding ventilation around electric vehicles being charged. This information still remains in the *NEC* today, as some vehicles may still be in service that require ventilation when charging.

Where the electric vehicle supply equipment (EVSE) is cord-connected, the location of the receptacle must be determined with **625.17** in mind. This can be difficult for new users of the *Code* to understand at first. The cord length provided by the manufacturer of the equipment will ultimately influence receptacle location as well.

While the location of the EVSE is not determined by the *NEC*, the location of the coupling means is. This is where understanding of the definitions of the article is essential. The *coupling means* is the connector used to connect the EVSE to the electric vehicle being charged. See **625.50** in the *NEC*.

The *NEC* continues to adapt to technology changes in the industry. Wireless charging of electric vehicles is also covered in **Article 625**, as are the requirements for using an electric vehicle as a source of back-up power for a premises.

Some electric vehicles can be charged using 460-volt DC electric vehicle supply equipment (EVSE).

ARTICLE 626, ELECTRIFIED TRUCK PARKING SPACES

Trucks that are parked for long periods of time or overnight at a truck stop may need a power supply in order to keep any perishable cargo refrigerated and to keep the sleeping area of the cabin comfortable for the driver. Modern electrified parking spaces can provide internet, video, and audio entertainment in addition to heating or cooling. Without an electrified parking space, these trucks would have to idle their engines to provide the power for the refrigeration and other power needs, burning fuel for the entire time they are parked.

The primary concern of **Article 626** is power. The amount of power needed for a set of electrified parking spaces is determined by a unique demand factor table. In other articles, the number of pieces of equipment or sites determines the demand factor applied; in **Article 626**, the United States Department of Agriculture (USDA) Plant Hardiness Zone Map is used instead. This map correlates with average temperatures and is used to determine cooling demand for refrigerated trailers. See the Informational Note below **Table 626.11(B)** in the *NEC* for more information. Note that these demand factors do not apply to the load calculation for refrigerated units that provide an environment for temperature-sensitive cargo.

ARTICLE 630, ELECTRIC WELDERS

When looking at **Article 630**, paying attention to the contents of the four parts of the article is important.

- **Part I**, as in other articles, provides the scope and general information.
- **Part II** is used when supplying arc welders. Arc welders use rod or wire that is consumed in the welding

process. An arc is struck between the rod or wire and the base materials to be welded together.
- **Part III** covers resistance welders, commonly called *spot welders*. Two pieces of metal are held together and electrodes are connected to the top and bottom. High current flows between the electrodes, welding the metal together.
- **Part IV** is concerned with welding cables. **Part IV** is used primarily for permanent welding installations where the welding cables are installed in dedicated cable tray, as covered in **630.42**.

Part II and **Part III** contain demand factor tables for duty cycles of welders. Welders are typically operated for short periods of time with short periods of rest between welding applications. Since the current is not flowing full time, the conductor ampacity can be reduced using the values in the tables. Make sure to use the table that corresponds to the type of welder being installed.

AUDIO EQUIPMENT, INFORMATION TECHNOLOGIES, AND ELECTRONIC EQUIPMENT

Articles 640, **645**, and **647** deal with applications where sensitive electronics may be involved.

The systems discussed in **Article 647** are frequently used with the audio systems covered in **Article 640**.

Article 645, Information Technology Equipment, contains special allowances and requirements for computer rooms.

Article 640, Audio Signal Processing, Amplification, and Reproduction Equipment

Article 640 covers permanent, temporary, and mobile applications of audio equipment. This includes recording systems, public address systems, and the like. See the Informational Note from **Article 640** for examples of systems covered.

Informational Note: Examples of permanently installed distributed audio system locations include, but are not limited to, restaurant, hotel, business office, commercial and retail sales environments, churches, and schools. Both portable and permanently installed equipment locations include, but are not limited to, residences, auditoriums, theaters, stadiums, and movie and television studios. Temporary installations include, but are not limited to, auditoriums, theaters, stadiums (which use both temporary and permanently installed systems), and outdoor events such as fairs, festivals, circuses, public events, and concerts.

There are a few notable exceptions to the scope of this article. For example, sound systems for theaters are not included in **Article 640**, as they are covered in **Article 520**. This article also does not include textual audible notification appliances for fire alarm circuits, which are covered in **Article 760, Fire Alarm Systems**.

Audio systems are typically perceived as a type of low-voltage system; however, this is not true of all audio systems, as some can produce significant voltages. The output side of some amplifiers will operate at 70 volts and can have signals peak at 300 volts. The perception of these systems being low-voltage may instead come from the allowed wiring methods. Class 1, 2, and 3 wiring methods, covered in detail in **Article 725**, are allowed to be used on the output side of an amplifier, according to **640.9**, where the amplifier is listed and marked to be used with the specific wiring method.

Although Class 2 or 3 methods may be allowed, they may not be mixed with other Class 2 or 3 circuits or with power-limited fire alarm circuits in a raceway or enclosure. *NEC* **725.139(F)** and **760.139(D)** do not allow the audio system circuits described in **640.9(C)** to share raceways. This is an important reminder for all Electrical Workers: when looking at a circuit, the source of the circuit determines the wiring, not the other way around. In this example, with audio circuits, it would be easy to assume that two circuits using Class 2 wiring methods may be located in the same raceway when that is not the case.

Articles 645 and 646, Information Technology Equipment and Modular Data Centers

Articles 645 and **646** cover rooms that are used for computer equipment. **Article 645** is for permanent IT equipment rooms, and **Article 646** covers modular data centers where the installation may be housed in a shipping container, but is intended for fixed installation.

An information technology equipment (ITE) room, as described in **Article 645**, typically is a room with a raised floor that allows the computer equipment within the room to be served from underneath. Simply adding a raised floor to a room does not make a room an ITE room, however. **Informational Note No. 2** directs the reader to *NFPA 75: Standard for the Protection of Information Technology Equipment* for more information.

The airspace beneath the raised floor of an ITE room is considered an environmental air space, or plenum. Unlike the installations addressed in **300.22**, however, many non-metallic wiring methods are allowed in these installations. In addition to the methods in **300.22(C)**, these methods are allowed:

- Rigid metal conduit
- Rigid nonmetallic conduit
- Intermediate metal conduit
- Electrical metallic tubing
- Electrical nonmetallic tubing
- Metal wireway
- Nonmetallic wireway
- Surface metal raceway with metal cover
- Surface nonmetallic raceway
- Flexible metal conduit
- Liquidtight flexible metal conduit
- Liquidtight flexible nonmetallic conduit
- Type MI cable
- Type MC cable
- Type AC cable
- Associated metallic and nonmetallic boxes or enclosures
- Type TC power and control tray cable

It is critical for the Electrical Worker to understand that the installation must be in compliance with **645.4** in order for all of these methods to be used. In a proper ITE room installation, any smoke produced from the nonmetallic wiring methods will stay confined to the ITE room and not be circulated to the rest of the building, which is the primary concern in **300.22(C)**. The underfloor space of ITE rooms must also have smoke detection for complete coverage and protection of the space.

Article 646, which covers modular data centers, has a similar layout and form to **Article 645**. The information technology equipment is to be installed within the enclosure, but support equipment, such as HVAC equipment, is permitted to be installed in another, separate enclosure. **Parts III** and **IV** of this article are particularly important to the Electrical Worker installing equipment within these enclosures, as these parts cover lighting and working space requirements.

Article 647, Sensitive Electronic Equipment

The purpose of **Article 647** is to provide the users of sensitive electronic systems in industrial and commercial applications with a means to reduce objectionable electrical noise on circuits supplying such systems. As **Article 647** does not define *sensitive equipment*, it is up to users to determine whether their needs require the use of this article and the installation of these balanced power systems.

Information related to this topic was previously located in **Article 530, Motion Picture and Television Studios and Similar Locations**, when the topic first appeared in 1996, but **Article 647** as a self-contained article was introduced in the 2002 edition of the *NEC*. A separate article devoted to sensitive equipment allows such systems to stand on their own from a *Code* perspective and to be no longer limited to the scope of **Article 530**.

The type of system discussed in **Article 647** is called a *balanced power system*. It is a separately derived system of 120 volts, single-phase, with 60 volts on each of two ungrounded conductors to an equipment grounding conductor. These ungrounded conductors have 120 volts between them. There are three requirements for the use of such systems:

- They must only be installed in commercial and industrial occupancies.
- Their use is restricted to areas under close supervision by qualified personnel.
- All the requirements of **Article 647** must be met.

Article 647 contains one of the only two voltage drop requirements in the *National Electrical Code*. It requires that the voltage drop on any of the branch circuits in a balanced power system be limited to 1.5%. If the branch circuit is supplying receptacle loads, the limit is only 1%. See **647.4(D)(1)** and **(2)**.

Receptacles fed by such systems need to be labeled according to **647.7**. **See Figure 6-2.**

Figure 6-2 Balanced Power System

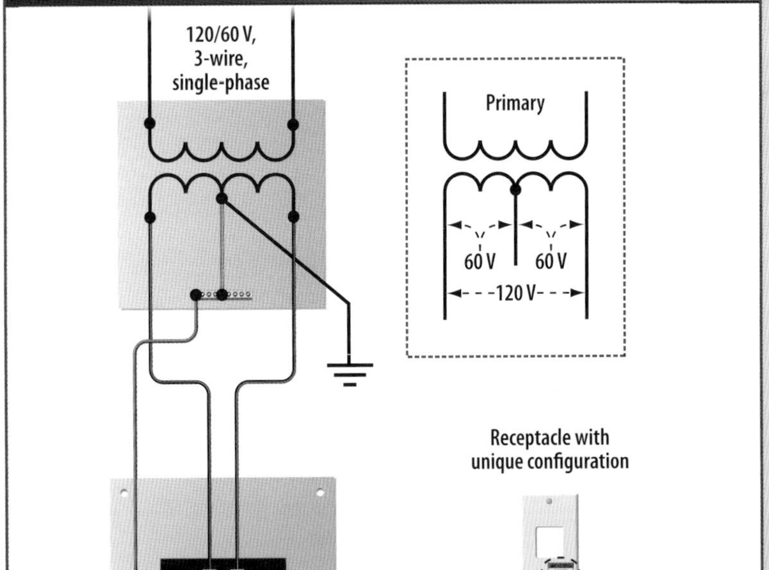

Figure 6-2. One use of balanced power systems is to eliminate noise relative to ground, making them an ideal choice for audio recording applications.

ARTICLE 650, PIPE ORGANS

Article 650 is one of the shorter articles in the *NEC*. This article has been in the *NEC* since the early 1900s, and went largely untouched until the 1990 edition of the *Code*. It was revised again in 2011 and 2014. Now **Article 650** includes modern wiring methods along with fiber optic cables. Circuits associated with pipe organs do not operate at more than 30 volts DC. **Article 250** also mentions pipe organs in **Section 250.112**.

INDUSTRIAL EQUIPMENT FOR MANUFACTURING AND PROCESSES

Articles 660 to **669** cover industrial equipment and equipment used for manufacturing or industrial processes.

Article 660, X-Ray Equipment

X-ray equipment that is not used in health care facilities is covered in **Article 660**. This includes x-ray equipment used at airports for security screening, x-ray machines used to inspect welds, and even x-ray equipment used to inspect artwork to detect counterfeits or forgeries. Important electrical safety concerns such as sizing and location of disconnecting means, conductor sizing, and control are addressed in this article.

Fact

Health care-related x-ray equipment is covered in **Part V** of **Article 517**.

Article 665, Induction and Dielectric Heating Equipment

The installation and wiring of dielectric and induction heating equipment used in scientific and industrial applications are covered in **Article 665** of the *NEC*. For conductive equipment that is being heated, the equipment is placed in a high-frequency magnetic field. For equipment with poor conductivity, the equipment is connected between electrodes.

Articles 668 and 669, Electrolytic and Electroplating Equipment

Electroplating is the process of covering a conductive object with a coating of metal, such as nickel or chrome. This process can be used, for example, to coat a bumper for a car. The bumper is submerged in a solution containing chromium or nickel, and a DC current is applied through the solution in the cell. The negative electrode is attached to the bumper, which attracts the metal in the solution to the bumper, thereby coating it.

Electroplating uses an electrolytic cell. The process of electrolysis in an electrolytic cell forces the decomposition of chemical compounds such as water into hydrogen and oxygen. It can be used to separate aluminum from bauxite, the principal source of aluminum, which also contains iron oxides and other compounds.

> **Fact**
>
> A battery is a type of electrolytic cell where the electrolyte is used as a source of electricity rather than electricity being pushed though the electrolyte. Storage batteries are covered in **Article 480** of the *NEC*. Also see **Section 706.8** for batteries used in energy storage systems.

Article 668, Electrolytic Cells, applies to electrolytic cells and related parts and equipment for the production of aluminum, cadmium, chlorine, copper, fluorine, hydrogen peroxide, and other compounds and elements according the scope of **Article 668**. While electrolytic cells can also be used for the production of hydrogen, this application is not within the scope of **Article 668**.

Article 669 not only covers equipment for electroplating, but also anodizing, electropolishing, and electrostripping. This last use is similar to electroplating, but the polarity is reversed, removing the outer coating from the object being treated.

Article 670, Industrial Machinery

NFPA 79: Electrical Standard for Industrial Machinery covers the connections and wiring within an industrial machine. Section 1.4 of *NFPA 79* refers to **Article 670** in the *NEC* for requirements regarding overcurrent protection and the conductors supplying the machinery.

The definition of *industrial machinery* in the *NEC* is rather broad, but it does not include portable tools or machines not fixed in place, nor does it cover machinery in dwelling units.

The Electrical Worker installing and connecting to a piece of industrial machinery must have a certain amount of information in order to complete the installation properly. **Section 670.3** contains the requirements for the requisite permanent nameplate, and details what information must be included on the nameplate.

In addition to being equipped with overcurrent protection and properly-sized conductors, industrial machinery must be protected within the limits of the machine's short-circuit current rating. Care must be taken to ensure the available fault current does not exceed the short-circuit current rating of the industrial machinery. See **Section 670.5**.

Article 675, Electrically Driven or Controlled Irrigation Machines

Anyone who has flown over the United States and looked below has seen the large green circles created by center-pivot irrigation machines. This specialized equipment comes with its own unique requirements, which are addressed in **Article 675** of the *NEC*. Much of the article is based upon **Article 430, Motors, Motor Circuits, and Controllers**.

These machines are located far away from the electrical supply, and the machines themselves can be about a quarter of a mile long (though each machine is custom-made for the particular field in which it will be used). Because of these great distances, the equipment grounding conductor is required to be the same size as the largest supply conductor, whereas in **250.122**, the equipment grounding conductor is generally smaller than the supply conductors.

These machines are located outside, subject to a large range of temperatures, and the motors have varying amounts of use depending how far they are away from the center pivot. **Article 675** includes language that addresses these issues. **Article 675** also covers lateral motion machines that move linearly across a field. **See Figure 6-3.**

ARTICLE 680, SWIMMING POOLS

There are eight parts to **Article 680**, as there are many different topics covered within the scope of the article. **Article 680** not only applies to swimming pools (both storable and permanent), it also applies to hot tubs, spas, hydromassage bathtubs, fountains, and wading and decorative pools. Auxiliary equipment such as pumps, filters, motorized pool covers, and electrically powered pool lifts are also included.

There are a few considerations unique to these types of installations that must be considered. Besides the water, which is an obvious hazard whenever electricity is involved, chemicals used in pools can contribute to corrosive conditions, which over time can gradually damage or destroy materials or equipment.

Major topics related to a safe installation around swimming pools include, but are not limited to:

- Clearances
- Receptacle locations
- Equipotential bonding
- Voltage limitations
- GFCI protection

Section 680.9 in **Part I** of this article covers clearance requirements for both low and high voltages above swimming pools and structures around pools, such as observation stands and diving boards. Lower-voltage systems, such as communications and network-powered broadband communication systems, have their own clearances. These clearances prevent people from accessing the conductors and potentially energizing the pool or surrounding equipment. Note that this requirement is located in **Part I** and therefore applies to the rest of the article unless amended.

Section 680.22 discusses receptacle requirements and restrictions. Where a permanently-installed pool is installed, at least one receptacle must be installed to serve the pool area. This receptacle must be located at least six feet from the inside wall of the pool in order to prevent a radio or any other equipment from being put in the water while it is plugged into the receptacle. The receptacle also cannot be located more than 20 feet from the pool so that it can serve the pool area. If it is located farther away,

Figure 6-3 Lateral Motion Irrigation Machine

Figure 6-3. *Lateral motion irrigation machines are one type of equipment covered in* ***Article 675****.*

persons may use an extension cord to power their radio, which might then allow them to put the radio in the water while energized. Any additional receptacles that are installed also must be at least six feet away from the pool. See **680.22(A)(3)**.

As of the 2020 edition of the *NEC*, pool equipment rooms also require a receptacle.

Lighting outlets, luminaires, and ceiling-suspended paddle fans in pool areas are covered in **680.22(B)**. *NEC* **680.22(D)** requires other outlets, including telephone, data, or fire alarm outlets, to be at least 10 feet away from the inside walls of the pool.

Section 680.26 covers equipotential bonding in pool areas. So-called "stray voltages" can be hazardous to swimmers even at very low levels. Bonding all the parts and area around the pool, including the water, will reduce voltage potentials. The parts that must be bonded together are spelled out in **680.26(B)(1)** through **(B)(7)**, and the pool water itself must be bonded in accordance with **680.26(C)**. The bonding of the pool water is usually accomplished by the bonding of a ladder rail or handrail at the steps entering the pool, both of which have sufficient surface area to meet the requirements of **680.26(C)**.

Electrical Workers are often surprised to find that all metal parts within five feet of the pool must be bonded, even if they are not related to the pool installation. See **680.26(B)(7)**. While some exceptions may apply, metal piping, metal awnings, metal fences, and metal window or door frames must all be bonded. **See Figurc 6-4.**

The bonding requirements in **Section 680.26** are essential for the safety of those using the pool. Mistakes can be costly to correct. Understanding this section is critical.

One of the definitions in **Section 680.2** is the term *low voltage contact limit.* Equipment that operates within these values is generally allowed to be installed using less-strict requirements, for example without GFCI protection.

Low Voltage Contact Limit. A voltage not exceeding the following values:

(1) 15 volts (RMS) for sinusoidal ac

(2) 21.2 volts peak for nonsinusoidal ac

(3) 30 volts for continuous dc

(4) 12.4 volts peak for dc that is interrupted at a rate of 10 to 200 Hz

GFCI protection made its first appearance in the *NEC* in 1968 in the context of swimming pools, so it is not surprising to see GFCI requirements in **Article 680** today. There are now dozens of requirements related to this

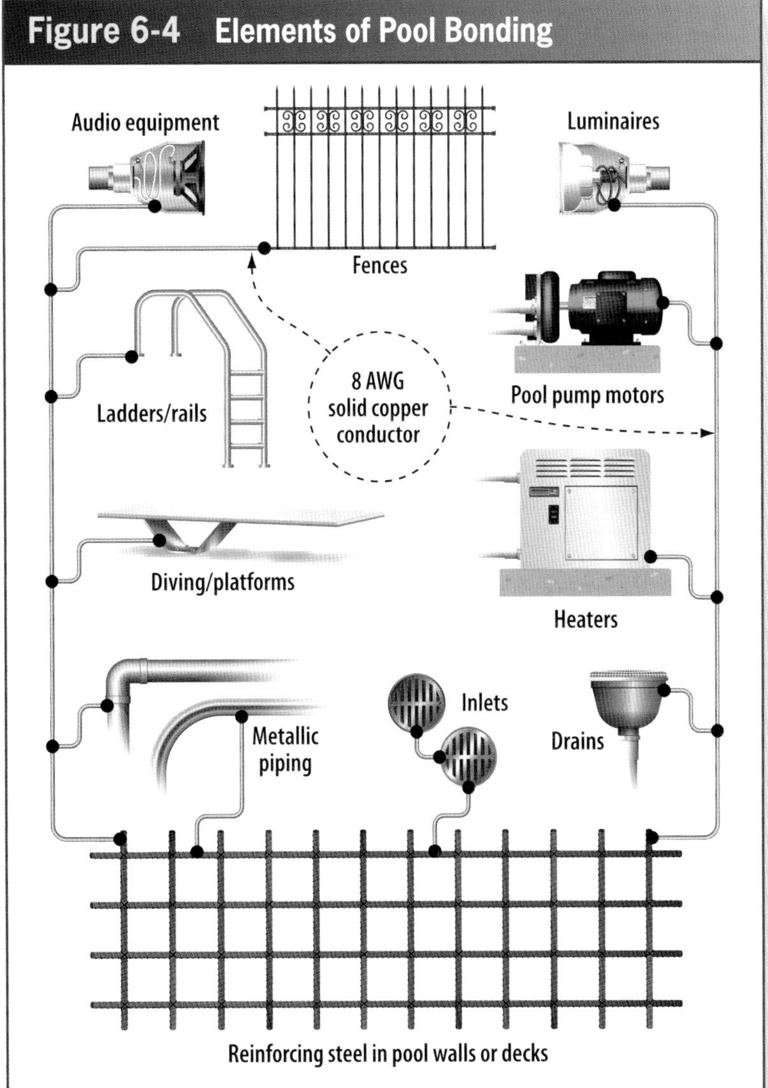

Figure 6-4. Bonding of the equipment and structures around a pool reduces voltage potentials in the area.

topic. GFCI protection may be required for the following:

- Receptacles
- Cord-connected equipment
- Heaters, pumps, and motors for most equipment
- Signs
- Lighting
- Motorized pool covers
- Audio equipment

Whether a specific device or piece of equipment requires GFCI protection may depend on:

- The type of pool, hot tub, or spa involved
- The device's distance or height from the pool, hot tub, or spa
- Indoor or outdoor location
- Voltage requirements

An installer must take special care in observing all relevant requirements for his or her particular installation.

ARTICLE 682, NATURAL AND ARTIFICIALLY MADE BODIES OF WATER

Bodies of water, whether pools or otherwise, bring a host of electrical shock-related issues that need to be managed. Land areas that are subject to tidal fluctuation, including areas along the ocean shore and portions of shoreline for rivers that end in the ocean; areas such as reservoirs that are subject to regular flooding; and areas subject to seasonal flooding from storms all carry the risk of water-related electrical shock if systems are improperly installed. Marinas, floating buildings, and pools have their own articles or sections in the *NEC*, but until the 2005 *NEC*, electrical installations around other bodies of water not covered in **Article 555** or **Article 680** were not specifically addressed. This changed with the introduction of **Article 682**.

Similar to **Section 555.3**, **Section 682.5** contains requirements for mounting equipment near various bodies of water. In addition to conforming to the general requirements in **210.8**, outlets fed by circuits up to 150 volts to ground and 60 amperes, single-phase, require GFCI protection. See **Section 682.15**. Feeders and branch circuits that are installed on piers require ground-fault protection that does not exceed 30 milliamperes. This is similar to the requirement found in **Article 555**.

ARTICLE 685, INTEGRATED ELECTRICAL SYSTEMS

An integrated electrical system in the context of **Article 685** is a system where orderly shutdown is required to prevent injury to persons or damage to equipment. The system must be maintained by qualified personnel.

A major requirement supplemented by **Article 685** is the location of overcurrent protective devices. Generally, **240.24(A)** requires overcurrent protection such as breakers and switches with fuses to be *readily accessible*. In **Section 685.10**, however, overcurrent devices are not required to be *readily accessible*, but may be *accessible*. This allows the devices to be mounted at heights to discourage or prevent operation by unqualified personnel.

Accessible (as applied to equipment). Capable of being reached for operation, renewal, and inspection.

Accessible, Readily (Readily Accessible). Capable of being reached quickly for operation, renewal, or inspections without requiring those to whom ready access is requisite to take actions such as to use tools (other than keys), to climb over or under, to remove obstacles, or to resort to portable ladders, and so forth.

ALTERNATIVE POWER SOURCES

Articles 690 and **691** cover photovoltaic (PV) systems, and **Articles 692** and **694** cover fuel cell systems and wind electric systems, respectively. Note that when any of these systems are installed, **Article 705, Interconnected Electric Power Production Sources**, should also be referenced.

Article 690, Solar Photovoltaic (PV) Systems

Generators can be power sources for a building, but in most cases generators do not serve as the primary or only source of power to a building. Instead, they are often used as a source of back-up power. Alternate power sources in **Chapter 6** are also used as back-up sources in some cases, but more commonly they are grid-tied so the solar array or wind turbine supplements the power coming in from the utility.

> **Fact**
>
> **Article 445** covers requirements specific to generators.

Article 690 has several diagrams to aid the user in identifying the parts of a solar photovoltaic system. The definitions of each of these parts are included in **Section 690.2. See Figure 6-5.**

There are eight parts to **Article 690. Part II** covers circuit requirements, which include considerations for voltage and current of a solar installation. The maximum voltage allowed depends on the type of building the system is installed on or in. In the case where a PV system is not located on or in a building, the maximum voltage may be as high as 1,500 volts. See **Section 690.7** in the *NEC*.

The DC voltage output of a PV system varies inversely with temperature. According to **Table 690.7(A)**, in extreme cold, the voltage can be 25% greater than the module's open circuit voltage. This maximum voltage shall be used to specify conductors, working spaces, and equipment where voltage ratings and limitations are expressed. For systems that are 100 kilowatts or greater, a licensed professional electrical engineer must provide a stamped and documented PV design for the installation.

The amount of current that a solar module can produce increases with the amount of light, or solar irradiance, to which the module is exposed. This can vary with the position of the sun, the time of year, and the day-to-day weather. In some cases, a partly cloudy day can actually increase the amount of solar irradiance on a panel due to the refraction offered by clouds.

When the leads of a module are shorted under standard test conditions, the module produces the short-circuit rating provided by the manufacturer. The *NEC* considers the maximum current to be 125% of the short-circuit rating of the module, or the sum of any parallel-connected modules or panels. Conductors must have an ampacity rating of 125% of this maximum current. See **690.8(A)** and **(B)** for more requirements or other permitted methods.

Another consideration when installing a PV system on a building has to do

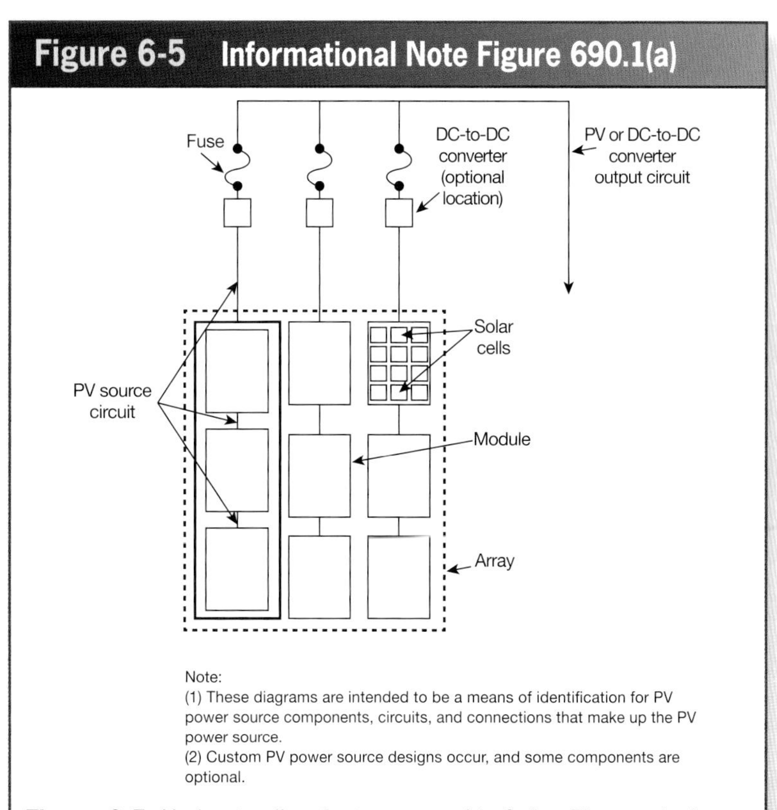

Figure 6-5. *Understanding the terms used in **Solar Photovoltaic (PV) Systems** is essential for the successful installation of these systems and understanding **Article 690**.*

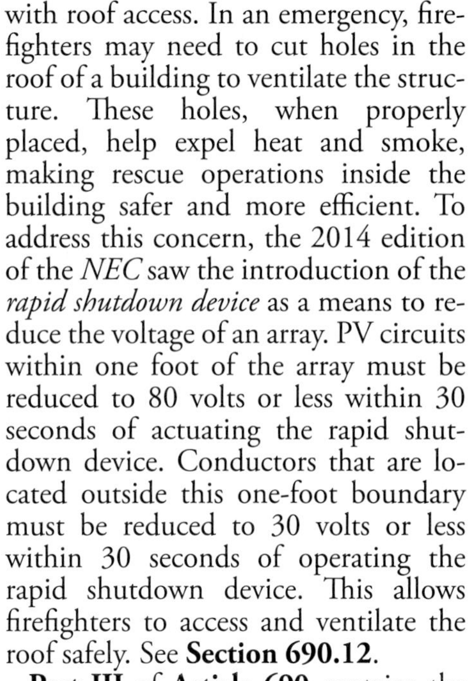

with roof access. In an emergency, firefighters may need to cut holes in the roof of a building to ventilate the structure. These holes, when properly placed, help expel heat and smoke, making rescue operations inside the building safer and more efficient. To address this concern, the 2014 edition of the *NEC* saw the introduction of the *rapid shutdown device* as a means to reduce the voltage of an array. PV circuits within one foot of the array must be reduced to 80 volts or less within 30 seconds of actuating the rapid shutdown device. Conductors that are located outside this one-foot boundary must be reduced to 30 volts or less within 30 seconds of operating the rapid shutdown device. This allows firefighters to access and ventilate the roof safely. See **Section 690.12**.

For additional information, visit qr.njatcdb.org Item #5515

Part III of **Article 690** contains the requirements for the disconnecting means, and **Part IV** covers wiring systems. Note that photovoltaic systems output DC voltage until they reach the inverter; as such, all disconnecting means, wiring methods, and overcurrent devices need to be DC-rated. The DC conductors shall be identified at all terminations and splices by an approved means. Marking tape is a standard method. See **690.31(B)(1)** for other identification methods that may be used.

One of the major topics in **Part V** of **Article 690** is ground fault protection (GFP). This protection is essential since the source of the power for PV systems is the sun, and in the case of an overload the fault can continue indefinitely, causing fires. GFP devices disconnect the faulted conductors or stop the supply power from the output of the PV array. Note that small panels consisting of two modules or less do not require GFP.

Devices and racks used for mounting and bonding PV modules must be listed and identified specifically for bonding PV modules. Exposed metal parts of the module frames, equipment, and enclosures must all be connected to an equipment grounding conductor. This is required regardless of the system voltage. See **Section 690.43**. The size of equipment grounding conductor is determined by **250.122**, or by using an assumed overcurrent protective device rated according to the value determined in **690.9(B)**.

Grounding electrode systems are required in **Section 690.47**.

Article 691, Large-Scale Photovoltaic Electric Supply Stations

Photovoltaic facilities that are not under the exclusive control of an electric utility and that generate at least 5,000,000 watts of power fall under the scope of **Article 691**. Facilities of this size are often visible from space.

These facilities have requirements that are not found in **Article 690**, including:

- The electrical design must be done by an electrical engineer.
- The facility may be accessible only to authorized personnel.
- Access to the facility must be limited by fencing.
- These systems may not be installed on buildings.

Article 692, Fuel Cell Systems

Fuel cell systems have only been included in the *NEC* for a few *Code* cycles, but they have already garnered the reputation for reliability. In fact, they are considered so reliable that they are allowed as an alternative source of power for health care facilities, with some additional conditions applied.

A fuel cell uses a chemical reaction, rather than combustion, to produce electrical current. Fuel cell systems can be stand-alone or interactive with other power sources, including utility sources. When fuel cells are connected as an interactive system, as with all alternative sources of power, **Article 705** applies.

Article 694, Wind Electric Systems

Another source of renewable energy that has recently been included in the *NEC* is wind energy harvested by wind turbines and their related equipment.

These can vary in size, and utilities may restrict the use of large turbines for private use depending on the area. **See Figure 6-6.** Wind turbines can be designed such that the blades will tilt to maximize wind generation, or such that in extreme high winds the blades may be rotated to shed wind. This protects the turbine from overspeed conditions. Turbine shutdown requirements are located in **Section 694.23**.

Figure 6-6 Wind Turbine

Figure 6-6. *Wind turbines are often installed in open areas, such as agricultural fields.*

ARTICLE 695, FIRE PUMPS

Much of **Article 695** is extracted from *NFPA 20: Standard for the Installation of Stationary Pumps for Fire Protection.* **Article 695** first appeared in the 1996 edition of the *NEC*, making it easier for Electrical Workers to access the requirements for the electrical versions of these units.

Unlike other equipment in the *Code*, a fire pump is wired such that it must run only when called into service. If a fire pump does not work, the building it is protecting could burn down. To avoid this situation, a fire pump is designed to continue to operate under locked rotor current conditions indefinitely, sacrificing itself if necessary.

Large industrial facilities may have more than one fire pump. Notice the equipment is kept off of the floor per the requirements in the NEC.

To this end, **695.4(B)(2)** requires the overcurrent protective device for a fire pump to indefinitely carry the locked rotor current of the largest pump motor and the full-load current of the other pump motors along with any accessory equipment. If this method is not chosen, then the other alternative is to install an overcurrent protective device in an assembly that is listed for fire pump service and that complies with the list items below:

- The overcurrent protective device shall not open within two minutes at 600% of the full-load current of the fire pump motor(s).
- The overcurrent protective device shall not open with a re-start transient of 24 times the full-load current of the fire pump motor(s).
- The overcurrent protective device shall not open within 10 minutes at 300% of the full-load current of the fire pump motor(s).
- The trip point for circuit breakers shall not be field adjustable.

To ensure the pump motor starts under load, the voltage drop at the terminals of the fire pump controller must not be more than 15%. This is one of only two voltage drop requirements in the *NEC*. See **Section**

695.7. Furthermore, conductors that supply fire pumps must comply with the following:

- They must be routed completely independent of other wiring.
- They must serve only loads that are associated with the fire pump.
- They must be protected from damage that may result from structural failure, fire, or accidents.
- If the conductors are routed inside a building and outside the fire pump room, they must be installed using a method that provides a two-hour fire rating.

Article 695 has many requirements that are unusual compared to other standards in the *NEC*. However, once the intent—that the water must flow to the sprinkler system of the building under any circumstance—is understood, these requirements make sense.

SUMMARY

Chapter 6 of the *NEC* contains requirements for specific equipment. When an Electrical Worker has the task of installing any of the equipment in the 27 articles of **Chapter 6**, consulting the relevant article is essential. While each article in the chapter does modify in some part the other chapters of the *NEC*, unless otherwise stated, **Chapters 1** through **4** still apply to these types of special equipment.

Any mistake during the construction of a building can be costly. Mistakes made when installing specialized equipment add to that cost exponentially. For these reasons, the Electrical Worker needs to be aware of the contents of **Chapter 6**. Some will find themselves specializing in certain equipment, such as swimming pools.

QUIZ

Answer the questions below using a copy of the *NEC* and by reviewing Chapter 6 of this text.

1. **The momentary rating on the nameplate of an industrial x-ray machine is 32 amperes. This current will flow for no more than __?__.**
   a. 5 seconds
   b. 10 seconds
   c. 1 minute
   d. 10 minutes

2. **The voltage drop for a 120-volt branch circuit feeding sensitive electronic equipment receptacles shall not exceed __?__.**
   a. 1%
   b. 1.5%
   c. 2%
   d. 3%

3. **Which of the following wiring methods is not allowed in an elevator machine room?**
   a. EMT
   b. ENT
   c. FMC
   d. IMC

4. **A commercial building that is accessible by pedestrians must have at least __?__ sign branch circuit(s) rated at least __?__.**
   a. 1 / 15 A
   b. 1 / 20 A
   c. 2 / 20 A
   d. None of the above; a sign circuit is permitted, but is not required

5. **A disconnect for electric vehicle supply equipment (EVSE) with a rating of 80 amperes shall be __?__.**
   a. accessible
   b. lockable in the open position
   c. within 25' of the EVSE
   d. all of the above

6. **Cords that are used to interconnect office equipment must not exceed __?__.**
   a. 12"
   b. 18"
   c. 24"
   d. 36"

7. **__?__ is the standard for fire protection of information technology equipment.**
   a. *NFPA 70*
   b. *NFPA 72*
   c. *NFPA 75*
   d. *NFPA 812*

8. **Optical fiber cable installed in a pipe organ shall be installed in accordance with __?__.**
   a. **Article 770**
   b. **Part I** of **Article 770**
   c. **Part V** of **Article 770**
   d. **Parts I** and **V** of **Article 770**

9. **The scope of __?__ includes a system that requires orderly shutdown procedures to ensure safe operation.**
   a. **Article 669**
   b. **Article 670**
   c. **Article 675**
   d. **Article 685**

10. **A receptacle must be installed at least __?__ and no more than __?__ from the inside walls of a permanently-installed swimming pool.**
   a. 5' / 15'
   b. 6' / 15'
   c. 6' / 20'
   d. 10' / 20'

NEC Chapter 7: Special Conditions

The seventh chapter of the *NEC* deals with special conditions. This includes back-up power systems, low-voltage and remote control signaling systems, fire alarm systems, fiber optic systems, and others. **Chapters 1** through **4** apply to all of these installations, except where **Chapter 7** modifies or amends those requirements. If any of these systems are in a special occupancy covered in **Chapter 5**, or supply special equipment in **Chapter 6**, those requirements also apply.

Objectives

- » Recall the titles of the articles found in **Chapter 7** of the *NEC*.
- » Locate definitions applicable to **Chapter 7** of the *NEC*.
- » Understand the differences between the three systems in **Articles 700**, **701**, and **702**.
- » Apply articles, parts, and tables to identify specific *NEC* **Chapter 7** requirements.

Chapter 7

Table of Contents

CHAPTER 7 OF THE *NEC*

Chapter 7 of the *NEC* is concerned with special conditions or applications of electrical installations. Some of these applications are encountered more frequently than others, and as a result Electrical Workers may find themselves specializing in one or more of these installation types. Even if a condition has never been encountered before, having a basic knowledge of these key areas and being able to locate the relevant information is essential.

Figure 7-1 Classifying Systems

Classes of Systems	
Class	**Minimum Operating Time in Hours Without Refueling or Recharging**
0.083	5 minutes
0.25	15 minutes
2	2 hours
6	6 hours
48	48 hours
X	Other time, in hours, as required by application, code, or user
Types of Systems	
Type	**Time Before Power is Restored**
U	Uninterruptible (UPS systems)
10	10 seconds
60	60 seconds
120	120 seconds
M	Manually operated, no time limit
System Levels	
Level	**Where Installed**
1	Where failure of the equipment to perform could result in loss of human life or serious injuries. Other codes such as *NFPA 99: Health Care Facilities Code* provide specific requirements. Typical applications include: life safety illumination, fire alarm systems, elevators, public safety communication systems, hazardous industrial processes, and smoke removal systems.
2	Installed to supply power automatically to certain loads that are not classified as emergency systems. When power is interrupted to these loads, hazards may be created or impede rescue or fire fighters. Typical uses include heating and refrigeration systems, communications systems, sewage disposal, lighting, and ventilation systems.

Figure 7-1. NFPA 110 *classifies systems by level, type, and class.*

Emergency and Standby Power Systems

It is important to discern the difference between the three types of back-up power systems in **Articles 700**, **701**, and **702**. Many use the terms "emergency power" or "emergency system" when discussing any kind of back-up power systems. This can lead to confusion as to what wiring method to use in the various installations, and this confusion can lead to potentially unsafe installations or time- and resource-wasting efforts to comply with requirements that do not apply or exist.

NFPA 110: Standard for Emergency and Standby Power Systems classifies back-up power supply systems by *Class*, *Type*, and *Level*. **See Figure 7-1.**

Article 700 correlates with Level 1, Class 10 systems; **Article 701** correlates with Level 2, Class 60 systems; and **Article 702** correlates with Class M systems.

To further aid the user, **Articles 700**, **701**, and **702** have a parallel numbering layout. **See Figure 7-2.**

Note that **Articles 700** and **701** share other parallel numbering patterns that do not apply to **Article 702**.

Article 700, Emergency Systems

Article 700 covers systems that consist of circuits intended to supply power within 10 seconds of interruption of the primary power source. When the term *emergency system* is properly used, it is referring only to the systems, equipment, and wiring covered in **Article 700**.

Emergency systems are required to be regularly tested and maintained per **700.3**. The authority having jurisdiction (AHJ) must witness the test, and the tests are done periodically on a schedule approved by the AHJ. The test shall be under the maximum anticipated load, and a written record of all tests and maintenance shall be kept.

If the emergency system relies on only one alternate source of power, a means to switch to a temporary alternate source must be provided when that power source is disabled due to maintenance or repair. Several companies and some larger electrical contractors exist that can provide this service by bringing in large trailer-mounted generators or uninterruptible supply systems (UPSs).

Section 700.4 states that the emergency power system must have enough capacity to carry all of the loads in accordance with **Article 220** or another approved method. If load shedding is implemented, the emergency source is permitted to supply back-up power to other than emergency system circuits, but the emergency system circuits must be given highest priority. See **700.4** for more details.

Emergency power systems require signage to be placed at the service entrance equipment showing the location and type of the emergency power sources. In addition to these requirements found in **700.7**, the emergency circuits must be permanently marked in accordance with **700.10**. Receptacles that are supplied from an emergency system must be identified with a distinctive color. The color is not specified, but local jurisdictions may mandate a color or it may be specified by the user of the system. This color will typically be the same for all

Figure 7-2 Articles 700, 701, and 702

Section of Article	Topic of Section
7XX.3	Tests and Maintenance
7XX.4	Capacity and Rating
7XX.5	Transfer Equipment
7XX.6	Signals
7XX.7	Signs
7XX.10	Wiring

Figure 7-2. *Parallel numbering is used in these articles to aid the user in finding and comparing the requirements between the different back-up power systems.*

signage, receptacles, and other forms of visual identification of the circuit enclosures and raceways. This makes it clear to the Electrical Worker what system he or she is working on.

The wiring of the emergency power system must be kept *entirely independent* from all other wiring. This is to prevent non-emergency loads and faults in one system from accidentally influencing another system. There are locations where emergency power and normal power must coexist, such as within a transfer switch. These locations are listed in **700.10(B)(1)** through **(B)(5)**.

This wiring must also be located such that it is protected from vandalism, fire, ice, flooding, or other conditions that could damage the wiring. In larger assembly occupancies, high-rises, and some educational occupancies, there are additional protective measures that must be taken according to **700.10(D)(1)**. These protective measures align with other requirements found in building codes for these occupancies.

Emergency systems in **Article 700** are not to be confused with legally required standby systems. Legally required standby systems, though frequently and inaccurately called "emergency systems," are systems that are required by other codes, municipalities, or AHJs. These systems are covered in **Article 701**.

Articles 701 and 702, Standby Power Systems

Article 701, Legally Required Standby Systems, has less stringent requirements than those found in **Article 700, Emergency Systems**. These systems correlate with Level 2, Type 60 systems in *NFPA 110* and are used where a longer time delay—up to 60 seconds—is acceptable. Signage requirements are similar to those found in **Article 700**, but wiring methods are less restrictive. The wiring of legally required standby systems is permitted to occupy the same raceways, enclosures, and cables as the normal power wiring.

Optional standby systems, covered in **Article 702**, are, as the title states, optional. This article would apply to a generator used to supply a person's home or the back-up power used to supply a computer server farm. Such systems for data storage may be critical for the continuity of a business, but these concerns are not typically reflected in codes or standards. Business owners, however, will specify any requirements found in **Articles 700** or **701** applicable for their needs.

Special Power Systems

The need for specialized power has increased tremendously in the past few decades. From redundant power systems for industrial processes to the interconnection of utility power and photovoltaic systems for a home, all systems must have minimum standards for their safe application and use. Some facilities simply must stay operational throughout a crisis, such as hospitals, emergency command centers, and central communication hubs used during natural or human-caused events.

Article 705, Interconnected Electric Power Production Sources

Many of the requirements found in **Article 705** were at one time located in **Article 690, Solar Photovoltaic (PV) Systems**. Since the use of renewable energy in its various forms has been on the increase, **Article 705** was added to the *NEC* in 1987. The scope of this article is the interconnection of one or more electric power sources such as solar, wind, or fuel cell operating in parallel with a primary source such as utility power. The Electrical Worker needs to be able to disconnect all power sources from a circuit to safely work on the circuit. This need is the rationale behind the identification requirements in **705.10**. The worker must know what power sources are available in order to shut them all off with confidence.

These alternative sources are permitted to be connected on either the line side or on the load side of the service. When connecting on the load side, the output of the inverter is typically connected to an overcurrent protective device in the panel, and it supplies power to the building in concert with the utility. **Section 705.12** contains requirements for the proper sizing of the overcurrent protective devices in the system and the bus within the panel. When a power source such as a photovoltaic system is added to a building, modification of the service overcurrent protective device or the panel may be necessary. See **705.12(A)** through **(E)**.

Article 706, Energy Storage Systems

Many forms of energy storage are available and more are currently being developed. When part of an energy storage system over one kilowatt-hour is used to store and provide energy during normal building operations, **Article 706** applies. Energy storage can take the form of chemical systems using batteries or mechanical systems using, for example, pneumatics or flywheels.

Article 708, Critical Operations Power Systems (COPS)

Any governmental entity such as a city, state, or federal agency has buildings or systems that are important to the security, safety, and health of the public or where the interruption of power to these facilities cannot be tolerated for other reasons. Examples of such facilities would include 911 call centers, emergency command posts, or even certain designated health care facilities. **Article 708** covers the more restrictive requirements for these facilities, which include:

- Physical security of the system, including restricted access
- Periodic testing and maintenance
- Marking of enclosures, indicating that the circuits within are part of the COPS
- Keeping COPS wiring independent of wiring from other systems
- Protecting feeders from fire and physical damage
- Installing HVAC, fire alarm, security, emergency communications, and signaling systems in a metal wiring method

While not required, should a building owner deem it necessary for the continuity of the mission of the facility, some owners may choose to use **Article 708** as a guide for the installation of a more robust electrical infrastructure.

Article 710, Stand-Alone Systems, and Article 712, Direct Current Microgrids

The systems covered in **Articles 710** and **712** can supply power to an entire building, to an area of a building, or to designated electrical loads within a building.

Stand-alone systems in **Article 710** are not connected to the utility or other power production network. They can consist of different sources such as energy storage systems, generators, and solar photovoltaic sources.

A DC microgrid is usually considered a stand-alone system. DC microgrid systems may have multiple sources of DC power such as batteries, solar, wind, and fuel cells. These systems provide DC power or AC power through an inverter. See **Article 712**.

Article 720, Circuits and Equipment Operating at Less Than 50 Volts, and Article 725, Class 1, Class 2, and Class 3 Remote-Control, Signaling, and Power-Limited Circuits

Historically, **Article 720** was used for low-voltage circuits providing power to remote buildings, such as barns, using a generator or series-connected batteries. While voltage was generally kept to 32 volts or less, currents could be high under some conditions. Most low-voltage applications today are installed according to **Article 725**. **Section 720.2** points out installations less than 50 volts that are not covered by **Article 720**.

Article 725, which covers low-voltage and Class 1, 2, and 3 circuits, applies to security wiring, thermostat wiring, control wiring for motor controls or lighting, and other signaling and control circuits. Note that fire alarm systems are not included, as those are the subject of **Article 760**.

Sections of **Article 300** that apply to low-voltage and Class 1, 2, and 3 circuits are referenced in **725.3(A)** through **(P)**. **Section 725.3** states:

> In addition to the requirements of this article, circuits and equipment shall comply with the articles or sections listed in 725.3(A) through (P). Only those sections of Article 300 referenced in this article shall apply to Class 1, Class 2, and Class 3 circuits.

Any references to **Article 300** must be complied with only if **Article 725** references a section in **Article 300**.

Class 1, 2, and 3 Circuits

Circuits in the *NEC* are defined by their source. Within **Article 725**, there are four types of circuits:

- Class 1 power-limited – **725.41(A)**
- Class 1 non–power-limited – **725.41(B)**
- Class 2 power-limited – **725.121(A)**
- Class 3 power-limited – **725.121(A)**

Class 1, 2, and 3 circuits do not include branch-circuit power and lighting circuits.

Class 1 power-limited circuits have a maximum voltage of 30 volts and a power limitation of 1,000 volt-amperes. One advantage of Class 1 circuits is that they can share a raceway or cable assembly with power and lighting conductors when serving the same piece of equipment. This can be advantageous with lighting controls. See **725.48(B)**.

A typical use of a Class 1, non–power-limited circuit would be motor controls. Motor control circuits are fed by the line voltage on a motor starter or from a control transformer, and then protected by a lower-ampere fuse. Notice that in **725.41(B)**, the only limitation to these circuits is that they must be no more than 600 volts.

Class 2 and 3 circuits are both power-limited. Neither of these circuits produce a hazard in regards to starting a fire, but where Class 2 circuits do not present a shock hazard, Class 3 circuits do, and additional safeguards are specified for these circuits to provide shock protection. See **Article 100** for the definitions of Class 2 and Class 3 circuits. Typical uses for Class 2 circuits are thermostat wiring and doorbell circuits, while examples for Class 3 circuits would include security systems or Class 2 circuits installed in wet locations.

A common installation error with Class 2 and 3 circuits is installing them in the same raceway or enclosure as power and lighting. Since the cabling used is rated at 150 to 300 volts, workers commonly assume these circuits can be installed with, for example, a 120-volt, 20-ampere power circuit. While **300.3(C)(1)** allows conductors with an insulation rating equal to or higher than the highest voltage present to be in the same raceway or enclosure, this does not include Class 2 or 3 wiring. See **Informational Notes No. 1** and **2** in **300.3(C)(1)**. Generally speaking, Class 2 or 3 circuits cannot be combined with power and lighting conductors unless one of the conditions in **725.136(B)** through **(I)** applies.

Communication cable has been used to provide power and data to devices such as security cameras within the same cable assembly as Class 2 and Class 3 circuits. **Section 725.144**, introduced in the 2017 edition of the *NEC*, provides the requirements for these installations, along with a new type of cable: Low-Power, or Type LP, cable. Standard communication cable can also be used. The advantage of using LP cable is that derating is not required where there are 192 cables or fewer bundled together. The ampacity of the cable or the value in **Table 725.144** is permitted be used to find the ampacity of the conductors. With standard communication cable, such as a Category 5 cable, use of this table is required when bundling. See **725.2** for the definition of a cable bundle.

Note that **Section 300.22** is mentioned in **725.3**, and therefore must be complied with. **Section 300.22** contains requirements for plenum spaces. **See Figure 7-3.**

Section 725.154 is another important section for the Electrical Worker to be familiar with. **Table 725.154** shows

Figure 7-3 Plenum and Non-Plenum Space

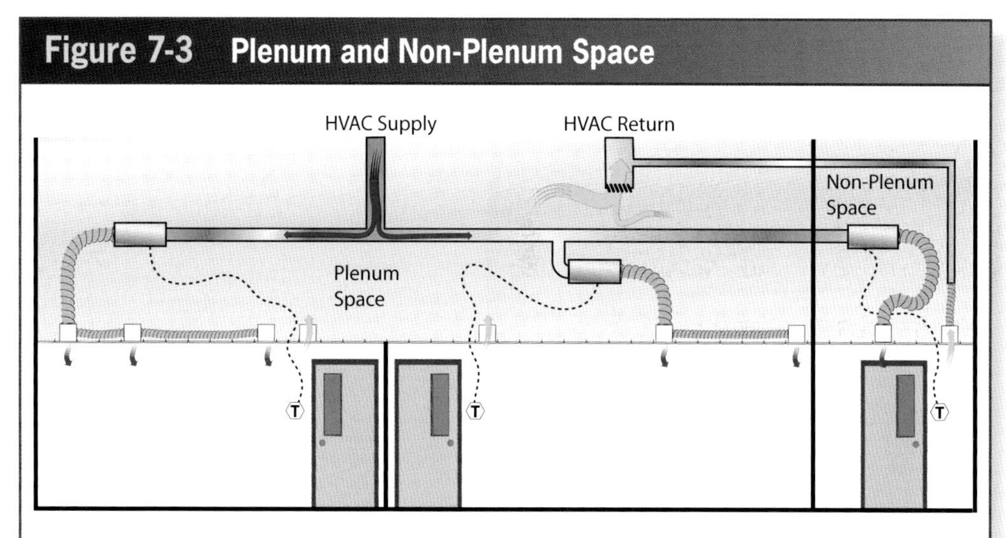

Figure 7-3. *The thermostat wiring above the room on the right would not require plenum-rated cabling, since the HVAC system serving this room does not use a common return. The ceiling space for the other two rooms is classified as a plenum space since the area is also used for purposes other than environmental air. See* ***300.22(C).***

the applications of the different types of low-voltage cabling assemblies according to their location within a building. **Figure 725.154(A)** in the *NEC* illustrates how one cable type can be substituted for another type depending on the rating of the cable. **See Figure 7-4.**

Figure 7-4 Figure 725.154(A)

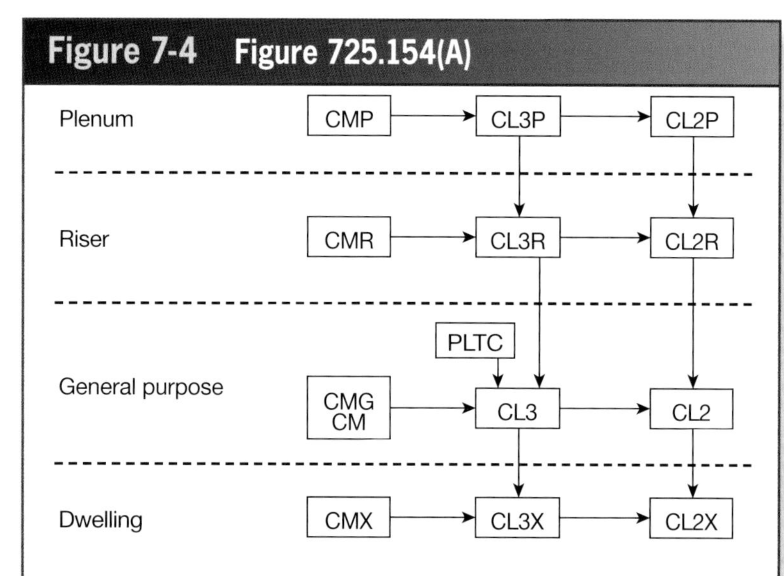

Figure 7-4. *The cable may be substituted in the direction of the arrows in the figure. For example, CMR cable may be substituted for CL3R.*

Articles 727 and 728, Special Wiring Conditions

Instrumentation Tray Cable: Type ITC in **Article 727** is used in industrial applications and may be installed in raceways or in cable tray. Instrumentation tray cable (ITC) is limited to being used in circuits with a maximum voltage of 150 volts and 5 amperes, although its voltage rating is 300 volts. Note that power and control tray cable, Type TC, is covered in **Article 336.** This is different than ITC cable and has a 600 volt rating.

There are many circuits that require protection from fire in the *NEC*. Circuits that need special protection are found in the following articles:

- **695, Fire Pumps**
- **700, Emergency Systems**
- **708, Critical Operation Power Systems (COPS)**
- **725, Class 1, Class 2, and Class 3 Remote-Control, Signaling, and Power-Limited Circuits**
- **760, Fire Alarm Systems**
- **770, Optical Fiber Cables**
- **805, Communication Circuits**

Whenever a circuit is installed for survivability of critical circuits, **Article 728, Fire-Resistive Cable Systems**, comes into play. Electrical circuit protective systems must be installed exactly as listed. Circuit integrity cables typically need to be supported every two feet, as that is how they are tested to achieve their listing. Where cables are required to be in a raceway, the raceway is typically supported every four to six feet, depending on the system. Connectors, couplings, vertical or horizontal orientations, and even the brand of raceway must all align with the listing requirements.

Perhaps the most important section of **Article 728** from an installer's point of view is **728.120**. Per this section, fire resistive cables are required to be marked with the system identifier. This identifier can be used to identify the proper way to install the system within the requirements of the listing.

For additional information, visit qr.njatcdb.org Item #5517

Article 760, Fire Alarm Systems

Fact

Most people do not have a fire alarm system in their home. The large majority of homes have what are called single-station or multi-station smoke alarms. These devices do not require a fire alarm panel to operate and are not within the scope of **Article 760**.

Many of the requirements in **Article 760** are also included in **Article 725**, since both articles are concerned with systems that are typically lower-voltage systems. As such, **Article 760** shares a parallel numbering structure with **Article 725**.

The article on fire alarm systems is broken down into four parts. **Part I** is general and applies to all fire alarm systems. **Section 760.3** lists what sections in **Article 300** apply to fire alarm systems. **Section 760.24** includes the requirements for installation. Note that while only **300.4(D)** applies in **725.24** due to the critical nature of fire alarm systems, all of **Section 300.4** applies within **760.24**.

While it is common for junction boxes containing fire alarm conductors to be identified with red paint in order to comply with **760.30**, the color red is not required. However, the box must be identified in some way to prevent unintentional signals on the fire alarm system.

In most jurisdictions, fire alarm systems must also be installed according to *NFPA 72: The National Fire Alarm and Signaling Code.*

Pull stations are located within five feet from the exit of a floor per NFPA 72. *Covers discourage malicious activation of the manual alarm device.*

Pull stations must be located where they are conspicuous, unobstructed, and accessible. A blocked pull station does not comply with these requirements from NFPA 72, *Chapter 17.*

Part II of **Article 760** covers non–power-limited fire alarm circuits. These circuits are similar to Class 1, non–power-limited circuits.

Fact

Not all fire alarm systems are low voltage! There are still existing fire alarm systems that are 120 volts. They consist of pull stations and bells. Older fire alarm systems had 110-volt notification appliance circuits. The devices were rated for lower voltage and connected in series. Large resistors in the panel provided the voltage drop needed when fewer than the maximum number of devices allowed were connected to the circuit. Make sure to verify the voltage before working on any system.

Part III, Power-Limited Fire Alarm Systems, is used more frequently with modern fire alarm systems, which operate at 24 volts or less. A major, often-overlooked difference with fire alarm systems compared to other systems in **Article 725** is that splices or terminations must be made using listed fittings, boxes, fire alarm devices, or utilization equipment. See **760.130(B)(1)**. Power-limited fire alarm system circuits cannot generally be installed in the same raceway or enclosure as power, lighting, and non–power-limited sources. See **760.136**. As was the case with Class 1, 2, and 3 cables, **Section 760.159** has requirements for different cable types and installation locations.

Part IV contains the listing and marking requirements of power-limited and non–power-limited fire alarm cable. It is important to note that while the *NEC* allows conductors as small as 26 AWG to be used per **760.179(B)**, manufacturer's instructions are also required to be followed. See **110.3(B)**. Manufacturers will have minimum conductor size and maximum conductor length requirements in their documentation that must be followed.

Fire alarm systems are for life safety. Should it ever have to be used, a properly-installed system will save lives.

Article 770, Optical Fiber Cables

The parallel numbering system from the two previous articles carries over to **Article 770**, which has six parts. When fiber optic cable first appeared in the 1984 edition of the *NEC*, there was much discussion about whether it should be included at all, since it is inherently non-conductive. In these early stages, the 1984 edition focused on the cable as a fuel in terms of fire and smoke creation. At that time, the article only contained one part and eight sections: **770.1** to **770.8**.

Fiber optic cable today offers many advantages over copper signaling paths. Fiber can transmit over extreme distances with minimal attenuation of signal strength. Furthermore, using fiber on networks when leaving a building mitigates concerns stemming from lighting strikes and other voltage transients.

Part I of **Article 770** covers many of the same topics as **Articles 725** and **760**. Installations shall conform to **300.4** and **300.11**. As with other low-voltage and communication signaling paths, abandoned fiber optic cables need to be removed according to **770.25**. Fiber optic cables that are not terminated or identified for future use with a tag are defined as *abandoned*, and the accessible portions must be removed.

Fact

The requirement to remove abandoned cable has been in the *NEC* since 2002. As buildings age and communication technologies improve and the needs of tenants increase, millions, if not billions, of feet of cable have been added to buildings across the nation. These cables become fuels and, regardless of their rating, contribute to smoke production in the event of a fire. As cables have been added over the years without the removal of old cabling, the weight of the accumulated cable can also cause structural issues. The removal of abandoned cables is required in **Articles 725**, **760**, and **770**, and **Chapter 8** of the *NEC*.

Part II of **Article 770** pertains to fiber optic cables outside and entering buildings. A common misconception that occurs whenever an installer runs fiber optic or communication cable into a building is that it must be terminated within 50 feet of the point of entrance into the building. This is not the case for all cable types. See **Section 770.48**. Only those cables that are not listed must be terminated within 50 feet of entering a building. In order for the installer to know if a cable is listed or not, it must bear one of the markings from **Table 770.179. See Figure 7-5.**

Any cable that is installed outside must also be rated for the location. If an unlisted outside plant cable must be run into the building for distances greater than 50 feet, the point of entrance can be extended using IMC or RMC raceways.

Grounding methods are covered in **Part IV** of the article. Conductive non–current-carrying parts of fiber optic cables are required to be bonded or grounded in accordance with **770.100(A)** through **(D)**. These members of the cable assembly can include armor or, in rare cases, strength members. Armor is used to protect the fiber from rocky soil conditions, rodents, or inside applications where the cable is subject to personnel stepping on the cable underneath raised floors. Grounding is essential to electrical safety. **Part IV** of **Article 770** needs to be understood before bringing fiber to a building.

Once the optical fiber cable is in the building, **Part V** must be consulted. Raceway fill requirements do not apply to fiber optic cable unless the cable also contains electric light and power conductors. See **770.110(B)**. Another advantage of fiber over copper signaling pathways is that armored fiber cable is allowed to be installed in the same raceway as power and lighting according to **770.133(A)**.

Section 770.154 and **Table 770.154(a)** show allowed cable substitutions and indicate where the various listed cables may be installed in a building.

Figure 7-5 Table 770.179 Cable Markings

Table 770.179 Cable Markings

Cable Marking	Type
OFNP	Nonconductive optical fiber plenum cable
OFCP	Conductive optical fiber plenum cable
OFNR	Nonconductive optical fiber riser cable
OFCR	Conductive optical fiber riser cable
OFNG	Nonconductive optical fiber general-purpose cable
OFCG	Conductive optical fiber general-purpose cable
OFN	Nonconductive optical fiber general-purpose cable
OFC	Conductive optical fiber general-purpose cable

Figure 7-5. *For an optical fiber cable to be listed, it must be labeled with one of eight markings.*

SUMMARY

Chapter 7 of the *NEC* is for special conditions. These conditions include:

- Emergency systems
- Back-up power
- Interconnection of power sources for a building
- Low-voltage and signaling systems
- Fire resistive wiring
- Fire alarm systems
- Fiber optic cabling

Each of these applications requires special and unique considerations. Articles involving back-up power and articles covering low-voltage systems use parallel numbering to aid the user of the *NEC* in finding what he or she needs. Since these articles are located in **Chapter 7**, each one may reference any article in **Chapters 1** through **7** per **90.3**.

Knowledge of the information in this chapter is essential for the Electrical Worker to avoid costly mistakes and assure proper installation.

QUIZ

Answer the questions below using a copy of the *NEC* and by reviewing Chapter 7 of this text.

1. **What is the maximum time allowed after power is interrupted for an emergency system to provide power?**
   a. 0 seconds
   b. 10 seconds
   c. 30 seconds
   d. 60 seconds
2. **Which of the following cable types can be used in a riser?**
   a. CL2P
   b. CMP
   c. CMR
   d. All of the above
3. **The 125-volt, 15-ampere receptacles supplied from a COPS shall have an indicator light or illuminated face to indicate that the receptacle __?__.**
   a. has power
   b. is currently in use
   c. is fed by a COPS system
   d. may not be used at this time
4. **What is the maximum voltage rating of PLFA cable?**
   a. 50 V
   b. 150 V
   c. 300 V
   d. 600 V
5. **A homeowner wants to have a generator installed for emergencies when the power goes out. Which article would apply to this application?**
   a. **Article 700**
   b. **Article 701**
   c. **Article 702**
   d. None of the above
6. **An energy management system shall not override power to __?__.**
   a. convenience receptacles
   b. COPS
   c. HVAC systems
   d. optional standby systems
7. **A group of cables supplying power over ethernet is considered to be bundled when they are closely packed for distances of __?__ or more.**
   a. 24"
   b. 36"
   c. 40"
   d. 44"
8. **A cable containing optical fibers and current-carrying electrical conductors is defined as a(n) __?__ fiber cable.**
   a. composite optical
   b. conductive optical
   c. non-conductive optical
   d. optical
9. **Where Class 3 cables emerge from a raceway that is used for protection, __?__.**
   a. a bushing must be installed
   b. a fitting must be installed
   c. the cable must be terminated within 50'
   d. none of the above
10. **Fire resistive cable systems shall be supported __?__.**
    a. according to the listing of the system
    b. every 2'
    c. every 5'
    d. every 6'

NEC Chapter 8: Communications

Chapter 8 in the *NEC* stands apart from **Chapters 1** through **7**. The requirements in **Chapters 1** through **7** do not apply to this chapter, unless an article in **Chapter 8** specifically references those requirements.

Prior to the 2020 *NEC*, **Chapter 8** was divided into articles that each covered a different communication-oriented application, which led to the duplication of requirements held in common across applications. Some other chapters of the *NEC* begin with a governing article, such as **Article 300** in **Chapter 3**. **Article 300** contains general requirements that apply to the rest of the articles in **Chapter 3**, with each subsequent article adding the unique requirements for the particular type of raceway or wiring method covered in that article. When an Electrical Worker installs EMT, he or she must comply with the requirements in both **Article 300** and **Article 358, Electrical Metallic Tubing: Type EMT**. In similar fashion, **Article 500** applies to the other articles in **Chapter 5** that cover classified locations.

Following this model, in the 2020 edition of the *NEC*, **Chapter 8** was revised to include **Article 800** at the beginning of the chapter. **Article 800** covers the requirements that are held in common between four different articles in **Chapter 8**.

Objectives

- » Recall the titles of the articles found in **Chapter 8** of the *NEC*.
- » Locate definitions applicable to **Chapter 8** of the *NEC*.
- » Understand the requirements of **Article 800** and their relationship to **Articles 805** through **840**.
- » Apply articles, parts, and tables to identify specific *NEC* **Chapter 8** requirements.

Chapter 8

Table of Contents

ARTICLE 800, GENERAL REQUIREMENTS FOR COMMUNICATIONS SYSTEMS

The scope of **Article 800** explains that the article covers general requirements for communications systems in these articles:

- **Article 805, Communications Circuits**
- **Article 820, Community Antenna Television and Radio Distribution Systems**
- **Article 830, Network-Powered Broadband Communications Systems**
- **Article 840, Premises-Powered Broadband Communications Systems**

Article 810, Radio and Television Equipment, is excluded from the scope of **Article 800**.

Article 800 is divided into five parts:

- **Part I, General**
- **Part II, Wires and Cables Outside and Entering Buildings**
- **Part III, Grounding Methods**
- **Part IV, Installation Methods Within Buildings**
- **Part V, Listing Requirements**

These parts are important to emphasize since their inclusion in **Article 800** means that they apply to *all* communications systems unless explicitly modified in one of the four other articles covering these systems. This means that when determining the requirements for a communications system installation, two articles may need to be checked: **Article 800** and the article covering the system type. **See Figure 8-1.**

Furthermore, when a communications system uses fiber optic cable, **Article 770** also applies. See **800.3(E)**. Many of the requirements that are found in **Articles 725**, **760**, and **770** are also found in **Article 800**, with the same parallel numbering system for usability. **See Figure 8-2.**

Section 800.24 has informational notes pointing the user to other accepted industry practices used when installing communications systems. Considerations such as the maximum distance of a cable run, the amount of pulling tension allowed on the cabling type, and details on how the cable is to be supported are included in these references. Many installations done in the field follow these references. In this case, an Electrical Worker also needs to understand requirements outside the *NEC*.

Section 800.27 requires the temperature limitations of cables to be followed. While this requirement is also included in the other articles in **Chapter 7**, it is found under a different section number in those locations.

Figure 8-1 Communication Installation

Figure 8-1. *For an installation of a communications system,* ***Article 800*** *and* ***Article 805*** *must both be consulted.*

Photo courtesy of Christopher Hill

Figure 8-2 General Requirements for Communication Systems

Section	Summary of Requirement
800.3	What other articles in the *NEC* apply to communications systems
800.21	Panels, such as suspended ceiling panels, that allow access to equipment shall not be blocked by communication wires and cables
800.24	Cables are supported according to **300.4** and **300.11** in such a way so that the cable is not damaged. No distance between supports is specified
800.25	Accessible parts of abandoned cables must be removed
800.26	Firestopping requirements for wall and floor penetrations

Figure 8-2. *Parallel numbering increases the usability of the* Code.

Part II covers issues that are encountered when installing cabling systems outside of buildings. This includes properly installing cables on poles and maintaining appropriate clearances from service conductors, roofs, and lightning conductors.

Grounding methods are covered in **Part III**. The 2008 edition of the *NEC* introduced the intersystem bonding termination (IBT) in **250.94**. The intersystem bonding termination is only for the grounding of communications systems. A busbar with the minimum dimensions stated in this section and connected per **250.94(B)** may be used in lieu of an IBT, and also is allowed to be used for connection to systems other than communications. Bonding conductors and grounding electrode conductors of communications systems are not required to be smaller than 14 AWG or larger than 6 AWG copper. See **800.100** for grounding and bonding requirements.

Fact

Some sections of **Article 800** apply to Class 2 and 3 circuit conductors when those circuits extend beyond a building. See **Section 725.141**.

Once the system is brought into the building, **Part IV** applies. Any raceway covered in **Chapter 3** is allowed to be used for a communications system, but the raceway fill requirements of **Chapter 3** and **Chapter 9** do not apply in these systems. See **800.110**. Note that other industry standards may limit conduit fill, since the amount of pulling tension on a cable assembly is a concern during installation, and too much tension applied during installation could mean the cable will not perform as required.

Per **800.113(A)**, cables are required to be listed. Unlisted cables run in a building must generally be terminated within 50 feet from the point of entrance. See **Sections 770.48**, **805.48**, and **820.48**.

Section 800.182 explains the general listing categories: plenum, riser, and general. This section shows how each cable type is related to the spread of fire and smoke.

ARTICLES 805, 820, 830, AND 840

The parts in each article of **Chapter 8** align by intent for the most part. This is done to make the *NEC* more user-friendly. **See Figure 8-3.**

Figure 8-3 Parts of Articles 805 Through 840

Part	Article 805 Communications Circuits	Article 810* Radio and Television Equipment	Article 820 Community Antenna Television and Radio Distribution Equipment	Article 830 Network-Powered Broadband Communications Systems	Article 840 Premises-Powered Broadband Communications Systems
I	General	General	General	General	General
II	Wires and cables outside and entering buildings	Receiving equipment	Coaxial cables outside and entering buildings	Cables outside and entering buildings	Cables outside and entering buildings
III	Protection	Amateur and citizen band transmitting and receiving stations – antenna systems	Protection	Protection	Protection
IV	Installation methods within buildings	Interior installation transmitting stations	Grounding methods	Grounding methods	Grounding methods
V	Listing requirements		Installation methods within buildings	Installation methods within buildings	Installation methods within buildings
VI			Listing requirements	Listing requirements	Premises powering of communications equipment over communications cables
VII					Listing requirements

*__Article 810__ is not included under the scope of **Article 800**

Figure 8-3. *Many of the parts in* ***Articles 805*** *through* ***840*** *cover similar or identical topics.*

For example, when installing communications circuits inside a building, the Electrical Worker must reference **Articles 800** and **805**. **Part I** and **Part IV** of each article will apply to the installation. Specifically, the following sections in **Part IV** of both articles will apply:

Part IV of **Article 800**

- **Section 800.110, Raceways, Cable Routing Assemblies, and Cable Trays**
- **Section 800.113, Installation of Wires, Cables, Cable Routing Assemblies, and Communications Raceways**
- **Section 800.154, Applications of Listed Communications Wires, Cables, and Raceways, and Listed Cable Routing Assemblies**

- **Section 800.179, Plenum, Riser, General-Purpose, and Limited Use Cables**

Part IV of **Article 805**

- **Section 805.133, Installation of Communications Wires, Cables, and Equipment**
- **Section 805.154, Substitutions of Listed Communications Wires, Cables, and Raceways, and Listed Cable Routing Assemblies**
- **Section 805.156, Dwelling Unit Communications Outlet** (Note that this applies only for a dwelling unit installation)

The titles of these sections reveal the more general nature of **Article 800** and the more specific nature of **Article 805**.

If an Electrical Worker wants to install telephone circuits in electrical metallic tubing (EMT), he or she would look at **800.110(A)(1)** to find that the EMT must be installed according to all of the requirements in **Chapter 3** and to find that there are no raceway fill requirements in **800.110(B)** for communications circuits. Notice that **Chapter 3** must be brought in by reference, as it would not otherwise apply to the articles in **Chapter 8**, per **Section 90.3**.

Reminder

Although raceway fill requirements do not apply to communications circuits, other standards may apply that limit the number of conductors in a raceway.

Moving to **Article 805**, the worker finds in **805.133** that the cables must be isolated from some kinds of circuits; however, this requirement would not apply to this installation, since only telephone cables are being installed in the EMT.

Electrical Workers installing systems involving **Articles 820**, **830**, and **840** would follow a similar path to find the requirements by first consulting **Article 800**, then consulting the article that corresponds to the type of system being installed.

Article 805, Communications Circuits

The scope of **Article 805** is communications circuits such as telephone and local area network systems. A *communications circuit* is defined in **Article 800** as the circuit that extends service from the communications utility or service provider up to and including the customer's communications equipment. This connection can be electrical or non–electrical, as would be the case if the pathway was accomplished with optical fiber or wireless technology. Typical copper wiring would be four twisted pairs of conductors in a cable assembly. **See Figure 8-4.**

Section 805.47 in **Part II** covers underground installation requirements specific to communications cable. This frequently involves the transition from outside plant cable from the utility or service provider to the interior installation, which requires listed communications cable. See **Section 805.48**.

Communications cables that are run outside of a building, whether aerial or underground, where exposed to accidental contact with electrical light and power conductors or lightning must be protected in accordance with **Section 805.90**. The primary protector shields

Figure 8-4 Typical Communications Cable Assembly

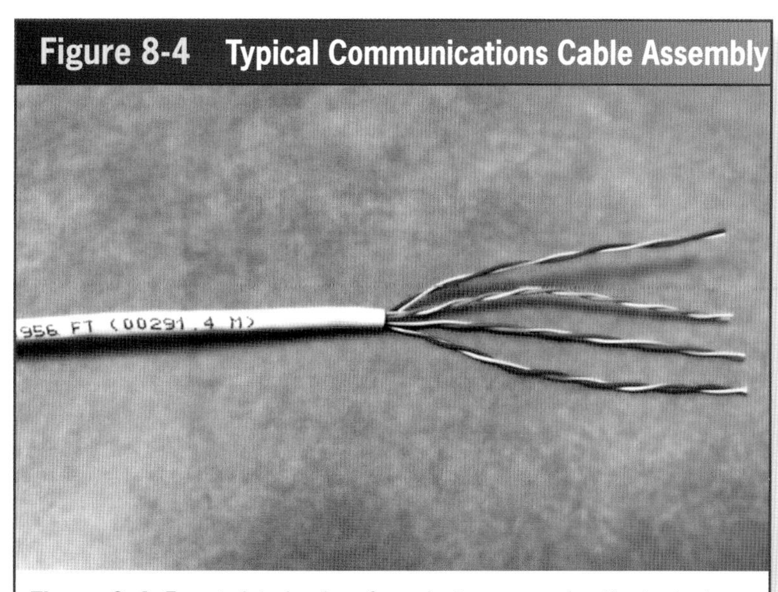

Figure 8-4. *Four twisted pairs of conductors comprise the typical communications cable.*

the interior wiring from exposure to lightning and other electrical systems. **See Figure 8-5.** Where communication circuits require primary protection according to **805.90**, they must be installed according to **Section 805.50**. To avoid the need for this protection, some system designers choose to use optical fiber cable between buildings. This is a common choice in areas where lightning strikes are likely to occur.

Once inside the building, the cables are routed in ceilings, raceways, or cable routing assemblies. **Section 805.133** contains the requirements for these installations. Communications circuits may be routed with the following:

- Class 2 and 3 circuits
- Power-limited fire alarm circuits
- Optical fiber cables
- Community antenna and radio distribution systems
- Low-power network-powered broadband communications circuits

Electric light and power conductors are not generally allowed in the same raceway or enclosure as communications circuits.

The requirements listed here are only a summary of the requirements contained in **805.133**. See the full text in the *NEC* for a complete list of requirements.

Figure 8-5 Primary Protectors

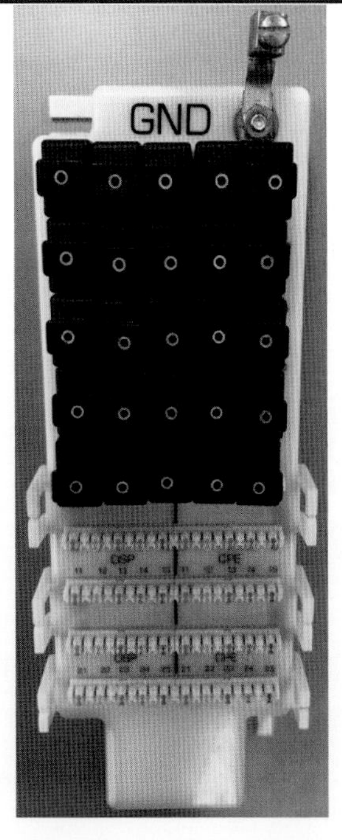

Figure 8-5. *Each black box on a primary protector protects a corresponding pair of conductors on the bottom part of the unit.*

Fact

Some sections of **Article 805** apply to fire alarm and Class 2 and 3 circuits leaving a building. See **Sections 760.32** and **725.141**.

Article 820, Community Antenna Television and Radio Distribution Systems

Article 820 applies to community antenna television and radio distribution systems, commonly known as *CATV* or *cable television*. Once a cable television provider brings the service to the building, **Article 820** should be followed to properly route the coaxial cable within the premises. This includes coaxial cables coming from antennas such as parabolic dish antennas where the service is provided via satellite.

Section 820.44 contains the requirements for coaxial cable installations that come from outside into the building and are attached to the building. When attaching coaxial cable to a building, a four-inch separation from electric light and power conductors is required where the light and power conductors are not in a raceway or separated from the coaxial cable by a nonconductor, such as the sheath of a cable assembly.

Section 820.47 covers what is required when a coaxial cable enters a building from underground. Notice that the same requirement is found in other articles of **Chapter 8** to terminate outside plant cables within the first 50 feet of entering the building. See **820.48**.

The shield of coaxial cable entering a building shall be bonded or grounded according to **Section 800.100**. Protective devices are allowed as long as they do not interrupt the grounding system of the building. See **Section 820.100**.

Fact

Some sections of **Article 820** apply where Class 2 and Class 3 conductors extend beyond the building. See **Section 725.141**.

Article 830, Network-Powered Broadband Communications Systems

A typical telephone is powered from the phone utility. However, broadband communications devices such as routers, cable modems, and digital subscriber line (DSL) services are powered by the building electrical system. Some providers of broadband services also provide power over the same cable as the broadband information. The cable is terminated on a network interface unit (NIU) that breaks the data signal into the video, audio, and voice components that are served to the building. This is where **Article 830** applies.

Remember that **Article 800** also applies to these installations. See the informational note in **800.1** and **90.2(B)(4)**.

> Informational Note No. 1: See 90.2(B)(4) for installations of circuits and equipment that are not covered.

> **90.2(B) Not Covered.**
>
> ...
>
> (4) Installations of communications equipment under the exclusive control of communications utilities located outdoors or in building spaces used exclusively for such installations.

Article 830 is applied to an installation after the point where the utility has exclusive control. Areas where exclusive utility control exists include conductors up on the utility pole or the conductors run to an area inside the building where access is restricted to utility personnel. Output circuits from the NIU to the equipment being served are covered within the scope of **Article 830**.

Article 840, Premises-Powered Broadband Communications Systems

Article 840 covers installation where a communications service provider supplies data, voice, video, and other signal types to a network terminal. A key difference between **Article 830** and **Article 840** is that in **Article 840**, the network terminal is powered from the premises, not the service provider.

Prior to the 2017 edition of the *NEC*, this article was limited to optical fiber pathways from the network terminal. However, signal pathways using copper conductors are also commonly used, and **Article 840** clearly addresses these installations.

The network terminal breaks the incoming signal from the service provider in the component signals of data, voice, audio, and other services and distributes them on a premises communications circuit and/or a premises community antenna (CATV) circuit.

> **Premises Communications Circuit.** The circuit that extends voice, audio, video, data, interactive services, telegraph (except radio), and outside wiring for fire alarm and burglar alarm from the service provider's network terminal to the customer's communications equipment up to and including terminal equipment, such as a telephone, a fax machine, or an answering machine.

> **Premises Community Antenna Television (CATV) Circuit.** The circuit that extends community antenna television (CATV) systems for audio, video, data, and interactive services from the service provider's network terminal to the appropriate customer equipment.

These circuits are installed in accordance with the services provided. For example:

- Communications circuits are installed according to **Part IV** of **Article 805**.

- Community antenna television and radio distribution circuits within the building are installed according to **Part V** of **Article 820**.
- Installations of optical fiber cables are installed according to **Part V** of **Article 770**.
- Class 2 and Class 3 circuits are installed according to **Part III** of **Article 725**.
- Power-limited fire alarm circuits are installed according to **Part III** of **Article 760**.

See section **840.3** for these and other articles that apply to the output circuits from the network terminal.

ARTICLE 810, RADIO AND TELEVISION EQUIPMENT

Article 810 could be called the most isolated article in the *NEC*, since it is not only located in **Chapter 8**, which stands alone from the other chapters, but it also stands alone from the rest of the articles in **Chapter 8**, as it is not included under the scope of **Article 800**. **Article 810** also differs from the other articles in **Chapter 8** because it specifically states that the requirements of **Chapters 1** through **4** *do* apply, except where modified by **Part I** and **Part II** of **Article 640**.

Article 810 covers requirements for TV and radio antenna systems. This includes parabolic dish type, wire strung type, and other types that may be used for amateur shortwave radio, citizen band radio transmitting and receiving equipment, and TV reception. Notice that while **Article 810** uses antennas to communicate, the other articles in **Chapter 8** use cable of some kind. Where coaxial cables are used to connect the antennas, they shall be installed according to the requirements of **Article 820**. *Lead-in conductors*, which connect the antenna to the receiving equipment, must be protected by listed surge protection devices and grounded in accordance with **810.21(F)**. See **810.6**.

Antennas and cables must be properly supported. As with other communications conductors and equipment, they are not allowed to be supported by or attached to the electrical service mast. The lead-in conductors must also be kept away from electric light and power circuits. See **Section 810.13** for details.

SUMMARY

Chapter 8 is the stand-alone chapter in the *NEC*. Whenever an Electrical Worker installs one of these systems, he or she must be aware of **Article 800**, which applies generally to **Articles 805**, **820**, **830**, and **840**. **Article 810** is not dependent on **Article 800**. Furthermore, other articles in the *NEC* may apply, depending upon the specifics of the installation. The 8XX.3 section of every article shows what other articles in the *NEC* apply.

QUIZ

Answer the questions below using a copy of the *NEC* and by reviewing Chapter 8 of this text.

1. **In general, when an unlisted outside plant cable comes into the building, it must be terminated within __?__.**
   a. 25'
   b. 50'
   c. 75'
   d. 100'
2. **Which part of Article 100, Definitions, applies throughout Chapter 8 of the *NEC*?**
   a. **Part I**
   b. **Part II**
   c. **Part III**
   d. **Article 100** does not apply to **Chapter 8**
3. **For communications systems, the bonding conductor or the grounding electrode conductor shall not be smaller than __?__.**
   a. 14 AWG
   b. 12 AWG
   c. 10 AWG
   d. 8 AWG
4. **Raceway fill requirements of Chapters 3 and 9 shall apply to __?__.**
   a. antenna lead in conductors
   b. communication systems
   c. medium-powered broadband communications cables
   d. none of the above
5. **Where attached on buildings, coaxial cables shall have a separation of at least __?__ from electric power conductors, or shall be separated from the power conductors by a firmly fixed nonconductor.**
   a. 4"
   b. 6"
   c. 12"
   d. 24"
6. **When fiber optic cable is used for the output circuit of a network interface unit, what part of Article 770 applies?**
   a. **Part I**
   b. **Part II**
   c. **Part IV**
   d. All parts of **Article 770** apply
7. **Abandoned network premises-powered broadband cables must __?__.**
   a. be maintained
   b. be entirely removed
   c. be tagged for future use
   d. have the accessible portions removed
8. **The outdoor antenna conductors for a citizen band radio system are never required to be larger than __?__.**
   a. 14 AWG
   b. 12 AWG
   c. 10 AWG
   d. 8 AWG
9. **Article 800 applies to which of the following articles?**
   a. **Article 805**
   b. **Article 820**
   c. **Article 840**
   d. All of the above
10. **What chapters of the *NEC* apply to Chapter 8?**
    a. **Chapters 1** through **4**
    b. **Chapters 5** through **7**
    c. **Chapters 1** through **7**
    d. None of the above

NEC Chapter 9: Tables and the Informative Annexes

Chapter 9 of the *NEC* consists of 12 tables and is followed by 10 Informative Annexes. The tables in **Chapter 9** are applicable only where referenced in the *NEC*, while the Annexes are for informational use only and are not enforceable parts of the *Code*. These tables and Informative Annexes supply the *Code* user with data to calculate conduit fill, ampacity, minimum raceway bending radius, voltage drop, power source limitations, and torque tightening values. They also include product safety standards, types of building construction classifications, supervisory control and data acquisition (SCADA) systems applications, and Americans with Disabilities Act (ADA) requirements.

Objectives

- Recall the tables of **Chapter 9**.
- Understand the use and application of the Informative Annexes.

Chapter 9

Table of Contents

TABLES

As required in **Section 90.3**, **Chapter 9** of the *Code* contains tables that apply as referenced throughout the *NEC*. As such, they are extremely valuable tools for the *Code* user. A basic understanding of the types of tables in **Chapter 9** and where they are referenced for use within the *Code* is necessary for quick navigation to the correct table. **See Figure 9-1.**

Table 1, Percent of Cross Section of Conduit and Tubing for Conductors and Cables

Table 1 is the benchmark for all permissible combinations of conductors in conduit and is commonly referred to as the "conduit fill table." **Table 1** is short and to the point; it lists only three types of conductor installations and their respective permitted percentage of raceway fill. **See Figure 9-2.**

Table 1 is referenced in the *NEC* wherever raceway fill requirements exist. For example, when using rigid metal conduit, **Section 344.22** references the fill specified in **Chapter 9**, **Table 1**.

It is important to note when using these tables that *notes* are enforceable, while *informational notes* are not enforceable. However, although the informational notes are not enforceable by an authority having jurisdiction (AHJ), they are good recommendations and worthy of attention. For example, **Table 1, Informational Note No. 2** has valuable information that, if not considered, could ruin an installation.

Table 2, Radius of Conduit and Tubing Bends

Table 2 provides a uniform minimum requirement for bends in conduit (to prevent damage to raceways) and reduction of the internal area of conductors.

Figure 9-1 Tables in Chapter 9 of the *NEC*

Table	Information in the Table
Table 1	Allowable percent of cross-section of conduit and tubing for conductors and cables
Table 2	Radius of conduit and tubing bends
Table 4	Dimensions and percent area of conduit and tubing for conduit fill per **Table 1** of **Chapter 9**
Table 5	Dimensions of insulated conductors and fixture wires
Table 5A	Compact copper and aluminum building wire dimensions and cross-sectional areas
Table 8	Conductor properties and DC resistance values of conductors
Table 9	AC resistance and reactance for 600-volt cables, 3-phase, 60 Hz, 75°C (167°F) – three single conductors in conduit
Table 10	Conductor stranding
Table 11	Class 2 and 3, AC and DC power source limitations
Table 12	Power-limited fire alarm AC and DC power source limitations

Figure 9-1. *The tables of **Chapter 9** apply as referenced throughout the* NEC.

Table 2 is referenced in the raceway articles. For example, for cases using rigid metal conduit, **Section 344.24** references the bend requirements of **Chapter 9, Table 2. See Figure 9-3.**

Please note that in certain conduit installations, the job specifications may need a larger bend radius than required by the *NEC*. This depends on the type of conductors being installed. For example, medium-voltage cable can be damaged by too much sidewall pressure, and larger-radius bends reduce sidewall pressure in a cable pull. Similarly, in the case of fiber optic cables, larger-radius conduit bends prevent damage to the fiber and yield maximum signal strength.

Table 4, Dimensions and Percent Area of Conduit and Tubing

Table 4 is extremely useful for the *Code* user when determining conduit fill. Information provided by this table includes total internal area and permissible fill area for several applications. **Table 4** is referenced in **Note 6** to **Table 1**, making **Table 4** applicable wherever **Table 1** is referenced in the *NEC*. **See Figure 9-4.**

In recent editions of the *NEC*, **Table 4** has been reorganized so that the more frequently used percentages are on the left side of the table. The first column is for the typical installation of conductors in a raceway. The 60% column is used where there is a conduit or tubing nipple 24 inches or less between enclosures,

Figure 9-2 Table 1 of Chapter 9

Table 1 Percent of Cross Section of Conduit and Tubing for Conductors and Cables

Number of Conductors and/or Cables	Cross-Sectional Area (%)
1	53
2	31
Over 2	40

Informational Note No. 1: Table 1 is based on common conditions of proper cabling and alignment of conductors where the length of the pull and the number of bends are within reasonable limits. It should be recognized that, for certain conditions, a larger size conduit or a lesser conduit fill should be considered.

Informational Note No. 2: When pulling three conductors or cables into a raceway, if the ratio of the raceway (inside diameter) to the conductor or cable (outside diameter) is between 2.8 and 3.2, jamming can occur. While jamming can occur when pulling four or more conductors or cables into a raceway, the probability is very low.

***Figure 9-2. Table 1** of **Chapter 9** is used whenever conduit fill is calculated.*

Figure 9-3 Table 2 of Chapter 9

Table 2 Radius of Conduit and Tubing Bends

Conduit or Tubing Size		One Shot and Full Shoe Benders		Other Bends	
Metric Designator	Trade Size	mm	in.	mm	in.
16	½	101.6	4	101.6	4
21	¾	114.3	4½	127	5
27	1	146.05	5¾	152.4	6
35	1¼	184.15	7¼	203.2	8
41	1½	209.55	8¼	254	10
53	2	241.3	9½	304.8	12
63	2½	266.7	10½	381	15
78	3	330.2	13	457.2	18
91	3½	381	15	533.4	21
103	4	406.4	16	609.6	24
129	5	609.6	24	762	30
155	6	762	30	914.4	36

***Figure 9-3.** The NEC requires a minimum radius in raceway installations, although often job specifications require an additional increase in the minimum radius.*

Figure 9-4 Table 4 of Chapter 9

Table 4 Dimensions and Percent Area of Conduit and Tubing (Areas of Conduit or Tubing for the Combinations of Wires Permitted in Table 1, Chapter 9)

Article 358 — Electrical Metallic Tubing (EMT)

Metric Designator	Trade Size	Over 2 Wires 40%		60%		1 Wire 53%		2 Wires 31%		Nominal Internal Diameter		Total Area 100%	
		mm²	in.²	mm²	in.²	mm²	in.²	mm²	in.²	mm	in.	mm²	in.²
16	1/2	78	0.122	118	0.182	101	0.161	61	0.094	15.8	0.622	196	0.304
21	3/4	137	0.213	206	0.320	182	0.283	106	0.165	20.9	0.824	343	0.533
27	1	222	0.346	333	0.519	295	0.458	172	0.268	26.6	1.049	556	0.864
35	1 1/4	387	0.598	581	0.897	513	0.793	300	0.464	35.1	1.380	968	1.496
41	1 1/2	526	0.814	788	1.221	696	1.079	407	0.631	40.9	1.610	1314	2.036
53	2	866	1.342	1299	2.013	1147	1.778	671	1.040	52.5	2.067	2165	3.356
63	2 1/2	1513	2.343	2270	3.515	2005	3.105	1173	1.816	69.4	2.731	3783	5.858
78	3	2280	3.538	3421	5.307	3022	4.688	1767	2.742	85.2	3.356	5701	8.846
91	3 1/2	2980	4.618	4471	6.927	3949	6.119	2310	3.579	97.4	3.834	7451	11.545
103	4	3808	5.901	5712	8.852	5046	7.819	2951	4.573	110.1	4.334	9521	14.753

Article 362 — Electrical Nonmetallic Tubing (ENT)

Metric Designator	Trade Size	Over 2 Wires 40%		60%		1 Wire 53%		2 Wires 31%		Nominal Internal Diameter		Total Area 100%	
		mm²	in.²	mm²	in.²	mm²	in.²	mm²	in.²	mm	in.	mm²	in.²
16	1/2	73	0.114	110	0.171	97	0.151	57	0.088	15.3	0.602	184	0.285
21	3/4	131	0.203	197	0.305	174	0.269	102	0.157	20.4	0.804	328	0.508
27	1	215	0.333	322	0.499	284	0.441	166	0.258	26.1	1.029	537	0.832
35	1 1/4	375	0.581	562	0.872	497	0.770	291	0.450	34.5	1.36	937	1.453
41	1 1/2	512	0.794	769	1.191	679	1.052	397	0.616	40.4	1.59	1281	1.986
53	2	849	1.316	1274	1.975	1125	1.744	658	1.020	52	2.047	2123	3.291
63	2 1/2	—	—	—	—	—	—	—	—	—	—	—	—
78	3	—	—	—	—	—	—	—	—	—	—	—	—
91	3 1/2	—	—	—	—	—	—	—	—	—	—	—	—

Article 348 — Flexible Metal Conduit (FMC)

Metric Designator	Trade Size	Over 2 Wires 40%		60%		1 Wire 53%		2 Wires 31%		Nominal Internal Diameter		Total Area 100%	
		mm²	in.²	mm²	in.²	mm²	in.²	mm²	in.²	mm	in.	mm²	in.²
12	3/8	30	0.046	44	0.069	39	0.061	23	0.036	9.7	0.384	74	0.116
16	1/2	81	0.127	122	0.190	108	0.168	63	0.098	16.1	0.635	204	0.317
21	3/4	137	0.213	206	0.320	182	0.283	106	0.165	20.9	0.824	343	0.533
27	1	211	0.327	316	0.490	279	0.433	163	0.253	25.9	1.020	527	0.817
35	1 1/4	330	0.511	495	0.766	437	0.677	256	0.396	32.4	1.275	824	1.277
41	1 1/2	480	0.743	720	1.115	636	0.985	372	0.576	39.1	1.538	1201	1.858
53	2	843	1.307	1264	1.961	1117	1.732	653	1.013	51.8	2.040	2107	3.269
63	2 1/2	1267	1.963	1900	2.945	1678	2.602	982	1.522	63.5	2.500	3167	4.909
78	3	1824	2.827	2736	4.241	2417	3.746	1414	2.191	76.2	3.000	4560	7.069
91	3 1/2	2483	3.848	3724	5.773	3290	5.099	1924	2.983	88.9	3.500	6207	9.621
103	4	3243	5.027	4864	7.540	4297	6.660	2513	3.896	101.6	4.000	8107	12.566

Article 342 — Intermediate Metal Conduit (IMC)

Metric Designator	Trade Size	Over 2 Wires 40%		60%		1 Wire 53%		2 Wires 31%		Nominal Internal Diameter		Total Area 100%	
		mm²	in.²	mm²	in.²	mm²	in.²	mm²	in.²	mm	in.	mm²	in.²
12	3/8	—	—	—	—	—	—	—	—	—	—	—	—
16	1/2	89	0.137	133	0.205	117	0.181	69	0.106	16.8	0.660	222	0.342
21	3/4	151	0.235	226	0.352	200	0.311	117	0.182	21.9	0.864	377	0.586
27	1	248	0.384	372	0.575	329	0.508	192	0.297	28.1	1.105	620	0.959

(continues)

***Figure 9-4. Table 4** of **Chapter 9** (reproduced in part) contains the information to calculate conduit size when filled with conductors of varying size.*

per **Note 4** of **Table 1**. Whenever one conductor is in a raceway, as is the case with gas, tube, and oil (GTO) for a neon sign or when a single multi-conductor cable is installed per **Note 9** of **Table 4**, then the column for "1 Wire" (53%) is used. Frequently, when only two wires are needed for a circuit, a cable assembly is used rather than a raceway; therefore the column applying to installations with two wires in a raceway is less commonly-used, and is located toward the right side of the table.

Table 5, Dimensions of Insulated Conductors and Fixture Wires

Table 5 provides the dimensions of insulated conductors and fixture wires needed to determine permissible conduit fill. **Table 5** is required to be used together with **Table 4** to determine permissible combinations of conduit fill. **Table 5** is referenced in **Note 6** to **Table 1**, making **Table 5** applicable wherever **Table 1** is referenced in the *NEC*.

Table 5A, Compact Copper and Aluminum Building Wire Nominal Dimensions and Areas

Table 5A provides the dimensions of compact copper and aluminum building wire needed to determine permissible conduit fill. **Table 5A** is required to be used together with **Table 4** to determine permissible combinations of conduit fill. **Table 5A** is referenced in **Note 6** to **Table 1**, making **Table 5A** applicable wherever **Table 1** is referenced in the *NEC*.

Table 8, Conductor Properties

Table 8 provides conductor properties for all conductor sizes from 18 AWG to 2,000 kcmil. The information provided includes circular mil area for all AWG sizes, stranding, diameters, and DC resistance. **Table 8** is referenced in **Note 8** to **Table 1** for determining area for bare conductors, making **Table 8** applicable wherever **Table 1** is referenced in the *NEC*.

Table 8 is commonly used for voltage drop calculations where the loads are primarily resistance loads with minimal influence from inductive loads, such as motors. The table has resistance values for aluminum conductors and for coated and uncoated copper conductors.

Table 9, Alternating-Current Resistance and Reactance for 600-Volt Cables, 3-Phase, 60 Hz, 75°C (167°F) — Three Single Conductors in Conduit

Table 9 provides the resistance and impedance values necessary for determining proper conductor application where voltage drop or other calculations are required. In long run installations of conductors, voltage drop is directly proportional to the length of the conductor from the source to the load. In AC circuits, the impedance changes with power factor and the type of raceway used.

A good example of such an installation are the branch-circuit conductor runs to parking lot light fixtures, which can be several hundred feet. Using **Table 9** allows the user to calculate the increased conductor size in order to minimize the voltage drop to the lighting poles.

Table 10, Conductor Stranding

The purpose of stranded cables in power applications is primarily for flexibility. There are five classifications of conductor stranding: Class AA, A, B, C, and D. Class AA conductors are only a few strands, or practically solid conductors, whereas Class D includes the most finely stranded cables, such as welding cables. The amount of strands in a cable typically determines the sizes of the cable; the more strands, the larger the outside circular dimension of the conductor. **Table 10** sets the number of strands for Class B and Class C copper conductors and for Class B aluminum conductors.

There are various reasons why the *NEC* declares the maximum amount of strands. One reason is related to the termination of a high-count stranded cable in a termination lug. If the strands are too small for a particular termination lug, they will not be terminated properly. See **Section 110.14** in the *NEC*.

Fact

Coated copper conductors, addressed in **Table 8** of **Chapter 9**, are coated with a tin alloy protecting the conductors and are used where corrosion is a concern.

Tables 11 and 12, Class 2 and 3 Power Source Limitations and Power-Limited Fire Alarm Power Source Limitations

Tables 11(A) and **11(B)** and **Tables 12(A)** and **12(B)** contain the power limitations for low-voltage wiring. These tables were at one time located in **Articles 725** and **760**, respectively; however, since the specifications in these tables are of primary interest to manufacturers rather than installers, the tables were moved to **Chapter 9** in the 1996 edition of the *NEC.*

Tables 11(A) and **11(B)** contain the power source limitations for Class 2 and 3 wiring for AC and DC, respectively. **Tables 12(A)** and **12(B)** contain the power source specifications for power-limited fire alarm sources for AC and DC, respectively.

INFORMATIVE ANNEXES

As established in **Section 90.3**, the Informative Annexes are not part of the requirements of the *NEC* and are included for informational purposes only. A basic understanding of the types of annexes and the information they contain is necessary for the *Code* user to utilize this valuable information. **See Figure 9-5.**

Informative Annex A

Informative Annex A, Product Safety Standards, provides a list of UL, ANSI/ISA, and IEEE product safety standards used for the listing of products required to be listed in the *NEC*. These standards are very helpful for the installer, as they not only provide vital information for safe installations but also include sound installation practices to ensure maximum performance of the equipment.

Informative Annex B

Informative Annex B, Application Information for Ampacity Calculation, provides application information for many types of ampacity calculations, including those for conductors installed in electrical ducts.

Figure 9-5 Informative Annexes

Annex	Information in the Annex
A	Product safety standards
B	Application information for ampacity calculations
C	Dimensions and percent area of conduit and tubing for conduit fill per **Table 1** of **Chapter 9**
D	Examples of load calculations and cable tray fill
E	Types of construction
F	Critical operations power systems
G	Supervisory control and data acquisition (SCADA) systems
H	Administration and enforcement
I	Recommended tightening torque tables
J	ADA standards for accessible design

Figure 9-5. *The Informative Annexes are a great source of vital information for the installer but are not part of the* NEC *requirements.*

Informative Annex C

Informative Annex C, Conduit, Tubing, and Cable Tray Fill Tables for Conductors and Fixture Wires of the Same Size, is a useful aid to the user of the *Code* in determining conduit fill when all the conductors to be installed in a raceway are of the same size and type. At the beginning of **Annex C** is a table of contents for the annex, which can save time finding the correct table.

Notice there are two tables for each raceway type. One table is for regular stranded conductors and the other is for compact stranded conductors. The tables with (A) in their title are for compact conductors.

Although **Annex C** is informational only, it is referenced in **Note 1** to **Table 1**. Recently, this table has been reviewed so that the information in the table matches the results of the calculation requirements for conduit fill in **Table 1**.

When an installation calls for conductors of different sizes, the minimum size raceway must be calculated using **Tables 4** and **5** in **Chapter 9**. Note that the values in the tables of **Annex C** contain only the typical percentage fill values of 31%, 40%, and 53%. Other percentage values, such as 60% for conduit nipples and 25% for seals in hazardous locations, must be individually calculated.

Informative Annex D

Informative Annex D, Examples, is provided to aid the *Code* user when making certain calculations required by the *NEC*. The requirements for calculations are demonstrated in this annex in the form of examples. As such, this annex is a "must read" for all electrical installers.

Examples D1 through **D12** are load calculations for different kinds of occupancies and applications, while **Example D13** discusses cable tray fill.

Informative Annex E

Informative Annex E, Types of Construction, is provided to aid the *Code* user when determining any of the five types of building construction, all of which are summarized in **Annex E**. **See Figure 9-6.** Several tables are included in **Informative Annex E** detailing fire resistance ratings, the maximum number of stories (floors) per type of construction, and various cross-references. Generally, fire resistance decreases as the roman numeral of the construction type increases.

Fact

The individual strands of wire in a compact conductor are formed and shaped to virtually eliminate space between the strands, resulting in a conductor of smaller dimensions.

Figure 9-6 Summary of Building Construction Types

Informative Annex E Types of Construction

This informative annex is not a part of the requirements of this NFPA document but is included for informational purposes only.

Table E.1 contains the fire resistance rating, in hours, for Types I through V construction. The five different types of construction can be summarized briefly as follows (see also Table E.2):

Type I is a fire-resistive construction type. All structural elements and most interior elements are required to be noncombustible. Interior, nonbearing partitions are permitted to be 1 or 2 hour rated. For nearly all occupancy types, Type I construction can be of unlimited height.

Type II construction has three categories: fire-resistive, one-hour rated, and non-rated. The number of stories permitted for multifamily dwellings varies from two for non-rated and four for one-hour rated to 12 for fire-resistive construction.

Type III construction has two categories: one-hour rated and non-rated. Both categories require the structural framework and exterior walls to be of noncombustible material. One-hour rated construction requires all interior partitions to be one-hour rated. Non-rated construction allows nonbearing interior partitions to be of non-rated construction. The maximum permitted number of stories for multifamily dwellings and other structures is two for non-rated and four for one-hour rated.

Type IV is a single construction category that provides for heavy timber construction. Both the structural framework and the exterior walls are required to be noncombustible except that wood members of certain minimum sizes are allowed. This construction type is seldom used for multifamily dwellings but, if used, would be permitted to be four stories high.

Type V construction has two categories: one-hour rated and non-rated. One-hour rated construction requires a minimum of one-hour rated construction throughout the building. Non-rated construction allows non-rated interior partitions with certain restrictions. The maximum permitted number of stories for multifamily dwellings and other structures is two for non-rated and three for one-hour rated.

Figure 9-6. *Construction types are detailed in* ***Informative Annex E.***

Informative Annex F

The information in **Informative Annex F** is useful to the *Code* user when compliance with the commissioning requirements of **Article 708, Critical Operations Power Systems (COPS)**, is necessary.

Informative Annex G

The information in **Informative Annex G, Supervisory Control and Data Acquisition (SCADA)**, is useful when implementing a security control and data acquisition system that may be installed together with a critical operations power system described in **Article 708**. These guidelines are useful and can influence the installation requirements when followed.

Informative Annex H

Informative Annex H, Administration and Enforcement, is provided as a model set of administration and enforcement requirements that could be adopted by a governmental body along with the electrical installation requirements of the *NEC*.

Informative Annex I

Informative Annex I addresses torque. In the world of electrical installations, torque applications are always present. Today, torque applications include everything from switches and circuit breakers to connectors and busbar. Manufacturers typically have recommended torque requirements for proper installation. **Informative Annex I** lists recommended torque levels as defined in *UL Standard 486A-B*.

Remember that the manufacturer's torque recommendations supersede all other listed standards. *NEC* **110.3(B)** requires the manufacturer instructions to be followed, and **110.14(D)** requires the use of an approved torque tool for terminations. However, where manufacturers do not provide specifications, **Annex I** can be a useful reference.

Informative Annex J

Informative Annex J allows the *Code* user to properly consider electrical design constraints in the context of the *2010 ADA Standards for Accessible Design*. These specifications are used in buildings accessible to the public. They relate to the location of receptacles, switches, fire alarm pull stations, and so forth to ensure that they are accessible to the public.

SUMMARY

The twelve tables of **Chapter 9** of the *NEC* apply only where referenced in the *Code*. These tables are necessary when applying many of the provisions of the *NEC*. Certain applications, such as conduit fill calculations, may include the use of several tables, including **Table 1** and **Tables 4** through **8**.

The Informative Annexes are intended only to aid the user of the *NEC*. These Informative Annexes offer useful information on product standards, conduit fill, types of construction, cross-reference tables, the application of **Article 708**, and administration and enforcement. Additionally, through the use of examples, **Informative Annex D** helpfully illustrates some of the calculations required in the *Code*.

QUIZ

Answer the questions below using a copy of the *NEC* and by reviewing Chapter 9 of this text.

1. **The tables in Chapter 9 of the *NEC* apply __?__.**
   a. as referenced in the *NEC*
   b. at all times
   c. only in **Chapters 5**, **6**, and **7**
   d. wherever they are useful
2. **Which table is referenced in raceway articles for conduit fill?**
   a. **Table 1**
   b. **Table 2**
   c. **Table 3**
   d. **Table 4**
3. **When the *NEC* requires wiring methods, materials, and equipment to be listed, Informative Annex __?__ provides additional information on the product standards.**
   a. **A**
   b. **D**
   c. **F**
   d. **G**
4. **The Informative Annexes of the *NEC* are __?__.**
   a. applicable as referenced
   b. mandatory requirements
   c. not mandatory requirements
   d. used only for special equipment
5. **Table 10 covers conductor __?__.**
   a. ampacity
   b. insulations
   c. properties
   d. stranding
6. **How many 6 AWG THW compact conductors may be installed in a one-inch schedule 80 PVC conduit?**
   a. 3
   b. 4
   c. 5
   d. 6
7. **Where forward reach is unobstructed, what is the highest elevation a switch can be mounted according to ADA requirements?**
   a. 15"
   b. 48"
   c. 54"
   d. 58"
8. **What is the cross-sectional area of a bare 10 AWG solid uncoated copper conductor?**
   a. 0.011 in^2
   b. 0.008 in^2
   c. 5.26 in^2
   d. 6.76 in^2
9. **What is the cross-sectional area of a 10 AWG THW conductor?**
   a. 0.0243 in^2
   b. 0.176 in^2
   c. 4.470 in^2
   d. 15.68 in^2
10. **When using a full shoe bender, such as a hand bender, what is the minimum radius allowed on a one-inch EMT conduit?**
    a. 4.5"
    b. 5"
    c. 5.75"
    d. 6"

The *Codeology* Method

Electrical literacy is important to the *Code* user. Productivity and profitability are at risk when an inexperienced worker or designer unknowingly violates a provision of the *NEC* and must do the work over again. Furthermore, complying with requirements that do not exist in the *NEC*, but only exist due to a misunderstanding of the *Code*, may add unnecessary costs to an installation or introduce new hazards.

The purpose behind the *Codeology* method is to help any user of the *NEC* find the information he or she needs as efficiently as possible. The *Codeology* method is dependent upon the user having some knowledge of the information that is within the *Code*. When using *Codeology*, the user will look for keywords in a question that will act as starting points to find the information he or she is looking for. With experience and knowledge of what is in the *NEC*, keywords in a question will be easier to identify.

Using the *Code* book can be difficult and frustrating since it is not meant as a design or instruction manual for untrained persons, per **90.1(A)**. The process of learning the *NEC* at the start takes considerable time and effort, but results in mastering the content and use of the *Code*.

Remember that the *NEC* changes every three years, with thousands of public inputs processed by the *Code*-making panels that add, modify, or remove requirements to provide a standard for electrical safety. The user must stay on top of these changes throughout his or her career.

By this point, the user should be familiar with the way the *NEC* is structured and divided into chapters, articles, parts, and sections. The *Codeology* method will explore the use of the index, table of contents, and keywords, along with other concepts to help the user find the applicable *Code* section.

Objectives

» Understand **Section 90.3** and its importance to the *Codeology* method.

» Identify the major topics and keywords of each chapter of the *NEC*.

» Use keywords and topics to find information in the *NEC*.

Chapter 10

Table of Contents

THE TABLE OF CONTENTS AND THE INDEX

The index located in the back of the *Code* book provides a link to finding words, phrases, or topics in *Code* articles and sections relating to a topic. The index can be utilized by the user picking a keyword or topic and looking it up in the index. However, it is important not to depend on the index alone, as there are so many topics a user will encounter that the index in the back of the book could not hold them all.

As is often the case with other reference books, using the table of contents, combined with a knowledge of how the *Code* is laid out, is the best method of finding information within the *NEC*. The table of contents will show the article by title and will guide the user to the correct part of the article.

Figure 10-1 Figure 90.3 in the *NEC*

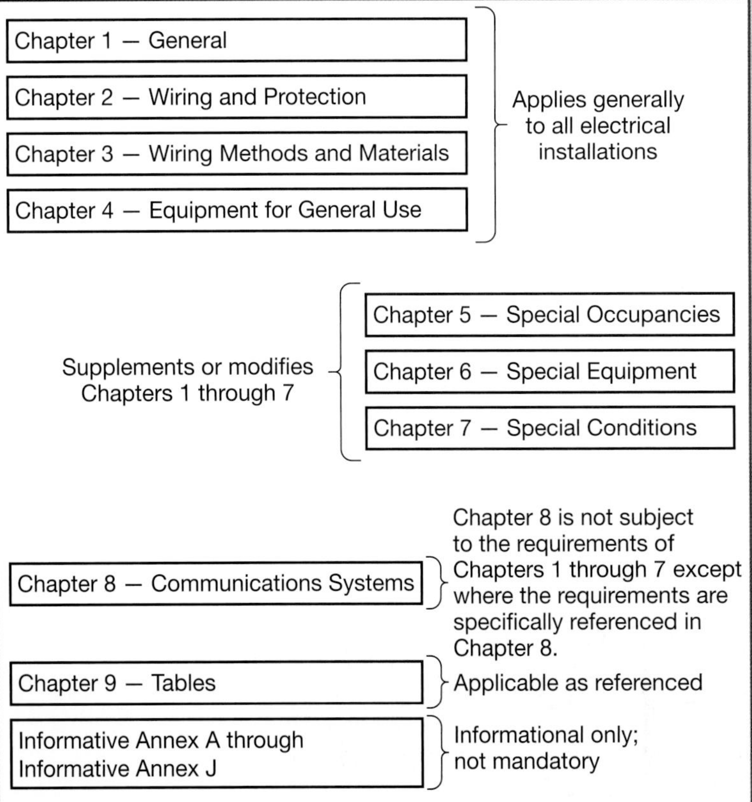

Chapter 1 — General
Chapter 2 — Wiring and Protection
Chapter 3 — Wiring Methods and Materials
Chapter 4 — Equipment for General Use
Applies generally to all electrical installations

Chapter 5 — Special Occupancies
Chapter 6 — Special Equipment
Chapter 7 — Special Conditions
Supplements or modifies Chapters 1 through 7

Chapter 8 — Communications Systems
Chapter 8 is not subject to the requirements of Chapters 1 through 7 except where the requirements are specifically referenced in Chapter 8.

Chapter 9 — Tables
Applicable as referenced

Informative Annex A through Informative Annex J
Informational only; not mandatory

***Figure 10-1. Figure 90.3** in **Article 90** is a useful guide for the* Codeology *method.*

The table of contents shows:

- The outline of the *NEC*
- The ten major sections of the *NEC*
- The subdivisions of each chapter into articles
- The subdivisions of each article into parts

THE IMPORTANCE OF ARTICLE 90

Article 90 is often overlooked as not very important, but it serves as the foundation of the *Codeology* method. Within the *Code*, there is a hierarchy of the chapters among each other, and therefore there is a priority system that should be used when looking for information in the *NEC*. **See Figure 10-1.** For example, **Chapter 8** is a stand-alone chapter and is not subject to the requirements in the rest of the chapters unless a section in **Chapter 8** specifically states otherwise and references a requirement in another chapter. Therefore, if a question deals with communications systems, **Chapter 8** should be consulted first, and in most cases is likely to be the only chapter consulted.

Similarly, **Chapters 5**, **6**, and **7** are to be given priority when a question arises that mentions a special occupancy type, special equipment, or special conditions. If multiple keywords are found that apply to more than one chapter, each one should be investigated.

Finally, if the question does not include a specific topic that is found in **Chapters 5** through **8**, attention should be given to the more general topics found in **Chapters 1** through **4**.

THE *CODEOLOGY* METHOD

With the *Codeology* method, each chapter is given a name to help the user know when to use that particular chapter. The names are as follows:

- **Chapter 1** - General
- **Chapter 2** - Plan
- **Chapter 3** - Build
- **Chapter 4** - Use
- **Chapters 5**, **6**, and **7** - The Specials
- **Chapter 8** - Communications
- **Chapter 9** - Tables

Chapters 1 through 4

Within the *Codeology* method, **Chapters 1** through **4** are organized according to how they are most likely to be used when applying the *NEC*: General, Plan, Build, and Use. Each of these chapters are associated with certain keywords or concepts.

Chapter 1

Chapter 1 is called the "General" chapter. It is applied generally to all installations and includes information such as general definitions, terminations, cooling and ventilation, and marking requirements that apply generally to all electrical equipment.

Working clearances are covered in ***Section 110.26*** *of the* NEC.

Keywords and concepts related to **Chapter 1** include:

- General definitions
- Examination, installation, and use of equipment (listed, labeled)
- Determination of approval
- Mounting and cooling of equipment
- Mechanical execution of work
- Electrical connections
- Flash protection
- Marking requirements general to systems or equipment
- Identification of disconnecting means
- Workspace clearance
- Manholes and tunnels
- General installation questions for over 1,000 volts

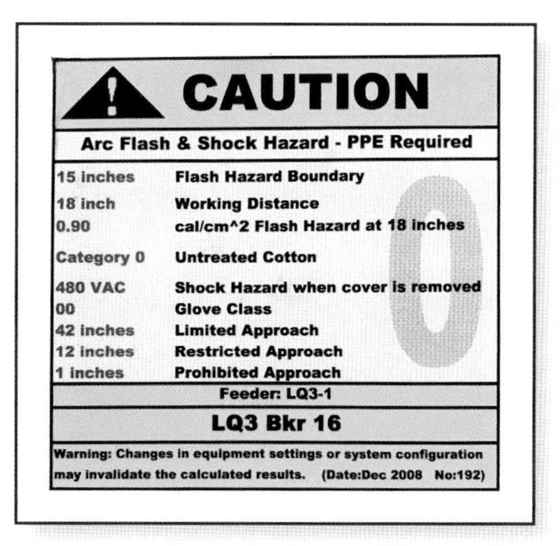

Labeling of equipment is covered in ***Article 110*** *of the* NEC.

Chapter 2

Chapter 2 is called the "Plan" chapter. This chapter deals with the planning of wiring, overcurrent protection, and overvoltage protection.

When an installation is being planned, the "big picture" items must be addressed first. *What circuits are needed? What are the loads that need to be planned for, and what calculations are needed? How will the service be installed and the electrical system be grounded?* These decisions are made at the beginning of the project, often before the first shovel touches the ground. Every user of the *Code* should be familiar with the topics of **Chapter 2**, as they are common to most installations.

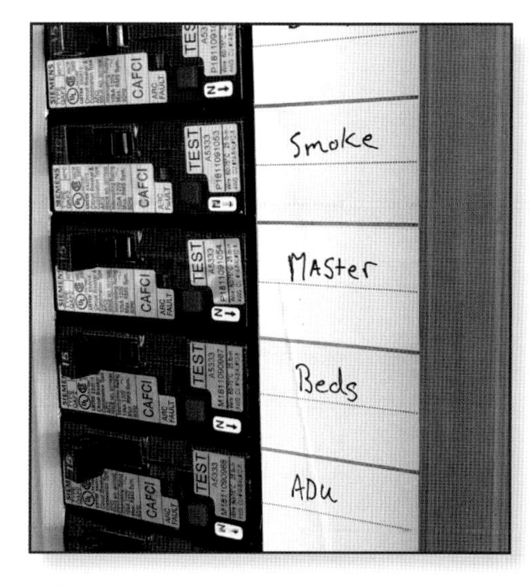

AFCI protection is the topic of ***Section 210.12.***

Keywords and concepts related to **Chapter 2** include:

- Use and identification of grounded conductors
- Branch circuits
  - Ground-fault circuit interrupters (GFCI)
  - Arc-fault circuit interrupters (AFCI)
  - Required receptacle outlets
- Feeder
- Service
- Load calculations
- Overcurrent protection
- Grounding
- Overvoltage protection
  - Surge arresters
  - Surge-protective devices (SPDs) 1,000 volts or less

*EMT raceways are covered in **Article 358** of **Chapter 3**.*

Photo courtesy of Ethan Miller

Chapter 3

Chapter 3 is called the "Build" chapter, and contains the requirements for performing the actual installation of the wiring system. Raceways, boxes, and conductors are all covered in **Chapter 3**. These items are the infastructure of the system that transports the electricity to where it is needed. When a question concerns wiring methods, boxes, or other considerations related to physically installing the system, think "Build" and consult **Chapter 3**.

Keywords and concepts related to **Chapter 3** include:

- General questions on wiring methods or materials
- General questions for conductors: uses permitted, ampacity, etc.
- Boxes of all types
- Conduit bodies
- Cabinets, meter sockets, wireways, etc.
- Installation, uses permitted, or construction of:
  - Specific kinds of raceways
  - Specific kinds of cable assemblies
  - Other electrical distribution methods
  - Support systems for wiring methods, such as cable tray and suspended ceiling power systems
  - Open wiring on insulators and outdoor overhead conductors

Chapter 4

Chapter 4 applies to the use of electricity and is called the "Use" chapter. When a question concerns equipment that controls, transforms, or uses electricity or is connected to the electrical system, think "Use" and look in **Chapter 4**. Luminaires (light fixtures), motors, transformers, batteries, capacitors, appliances, switches, and receptacles are among the many items covered in **Chapter 4**.

*Requirements for the use of listed countertop receptacles are found in **Section 406.5**.*

Keywords and concepts related to **Chapter 4** include:

- Cords or cables
- Fixture wires
- Switches

- Receptacles
- Switchboards and panelboards
- Industrial control panels
- Luminaires (lighting fixtures)
- Appliances
- Heating equipment
- Motors and controllers
- Air-conditioning equipment
- Generators
- Transformers
- Phase converters
- Capacitors, resistors, and reactors
- Batteries
- Electrical equipment over 1,000 volts

*The requirements for motor control centers are covered in **Article 430** of the NEC.*

Chapters 5 through 7

When a question or need arises that involves special occupancies, systems, or equipment, it is usually best to look at one of the special chapters first, unless the question is extremely general.

Chapter 5

Chapter 5 is titled **Special Occupancies**. This not only applies to entire buildings of a certain kind or use, but can also apply to different parts or areas of a building and different uses of an area or building. It is important to know what this chapter covers. When installing equipment in special occupancies, such as a hospital or classified (hazardous) location, mistakes can be costly.

*Dentist offices are defined in **517.2** and covered in **Article 517**.*

Keywords and concepts related to **Chapter 5** include:

- Hazardous (classified) locations
  - Combustible vapors, dusts, fibers, and flyings
  - Commercial garages and motor fuel dispensers
  - Spray booths and applications
  - Aircraft hangers
- Hospitals and all health care facilities
- Places of assembly
- Theaters
- Studios
- Carnivals
- Permanent amusement attractions
- Marinas
- Floating buildings
- Temporary power
- Manufactured buildings, motor homes, RVs, and RV parks
- Agricultural buildings and farms

***Article 555** needs to be consulted when working around marinas and floating buildings.*

Chapter 6

Chapter 6 is titled **Special Equipment**. Some equipment has particular requirements that may be more restrictive or less restrictive than is typical. Overcurrent protection, grounding, and conductor sizing are a few items that can vary depending on the equipment chosen.

Keywords and concepts related to **Chapter 6** include:

- Electric signs
- Cranes
- Elevators
- Electric welders
- Audio systems
- Information technology (IT) equipment
- X-ray equipment
- Swimming pools
- Solar photovoltaic systems
- Fuel cell systems
- Wind electric systems
- Fire pumps

Chapter 7

Chapter 7, the last of the special chapters, is titled **Special Conditions**. These conditions include specialized power such as energy management, fire alarm and fire-related wiring, low-voltage applications, and fiber optics.

Keywords and concepts related to **Chapter 7** include:

- Emergency systems
- Legally required and optional standby systems
- Systems with Class 1, 2, and 3 power sources
- Energy management systems
- Fire alarm systems
- Fire resistive cabling systems
- Fiber optics

Chapter 8

Chapter 8 has the requirements for communications systems, including telephony, data, radio and television, and CATV systems.

Keywords and concepts that are related to **Chapter 8** include:

- Communications circuits
- Radio and TV
- CATV systems
- Network-powered broadband systems
- Premises-powered broadband systems

With time and experience, these concepts and keywords will be committed to memory, enabling the Electrical Worker to efficiently find what he or she is looking for.

Chapter 9 and the Informative Annexes

Section 90.3 states that the tables in **Chapter 9** are applicable as referenced. The tables contain requirements regarding raceway wire fill limitations, wire sizes and resistances of conductors, and other information. These tables are referred to throughout the *Code* and are enforceable.

The Informative Annexes of the *NEC* are located at the end of the book. They provide helpful information, such as calculation examples, conduit fill for installations involving the conductors of the same size, and other information helpful to the application and installation of wiring systems.

USING THE *CODEOLOGY* METHOD

Section 90.3 gives a useful outline of how the *Code* is organized and also how to look for the answers to the questions a user has. While the *Code* itself is organized so that it starts with general applications and moves to more specific applications, it is recommended to look for the answers to a question by moving from specific to general. By doing this, the user of the *NEC* is more likely to find all of the requirements for the application.

Depending on the keywords in the question, look in the *NEC* in the following order:

1. **Chapter 8**, Communications
2. The Specials
   Chapter 7, Special Conditions
   Chapter 6, Special Equipment
   Chapter 5, Special Occupancies
3. General, Plan, Build, and/or Use
   Chapter 4, Use
   Chapter 3, Build
   Chapter 2, Plan
   Chapter 1, General

In most cases, checking all of these chapters is not necessary. Unless the question has keywords that are associated with the special chapters or communications, in most cases the Electrical Worker can begin with General, Plan, Build, or Use.

While Plan, Build, and Use are good guidelines for finding the required information, there is some crossover between some tasks. For example to "build" a branch circuit, **Article 210** in **Chapter 2** must first be consulted in order to "plan" the circuit. Once the requirements of **Article 210** are met, the building begins with the proper selection of the raceway, conductors, etc. in accordance with **Chapter 3**. Since the branch circuit may be used to feed equipment or devices such as lighting, receptacles, or a motor, these items may have additional requirements in **Chapter 4**.

For example, consider the question, "Can fuses or circuit breakers be installed in a dwelling unit bathroom?" Seeking a clue or keyword to begin the process, first determine that the question is asking for the location of *fuses and circuit breakers*, which is installation information. This question does not concern communications or any of the special chapters, so the user may move on to **Chapters 1** through **4**.

This is a "Plan" section question (**Chapter 2** in the *NEC*), but due to the fact it is very generic in nature, the next step is to look at the Table of Contents to locate the relevant section. **Article 240** addresses the subject of overcurrent protection, which is what fuses and circuit breakers are, and **Part II** of **Article 240** includes "Location" of overcurrent devices. **Section 240.21** is the first section regarding location.

Notice two things here: some basic knowledge is required, as the user must know that circuit breakers and fuses are types of overcurrent protection, and must be able to surmise that information about them will therefore be found in **Article 240**, generally. Furthermore, note that the question is about location, which is a highly general topic since all fuses and circuit breakers must have a location. Therefore this information is more likely to be found toward the first parts of the article.

Now scan the pages, reading the bold section titles only, looking for a clue to answer the question. **Section 240.24** contains the clue to answer the question: **240.24(E) Not Located in Bathrooms.** Within this section, the answer is found to be: "overcurrent devices shall not be installed in dwelling unit bathrooms."

240.24 Location in or on Premises.

...

(E) Not Located in Bathrooms. In dwelling units, dormitory units, and guest rooms or guest suites, overcurrent devices, other than supplementary overcurrent protection, shall not be located in bathrooms.

For another example that supports the need to understand terms, consider the question, "Can a panelboard be mounted in a clothes closet?" Immediately the keyword *panelboard* should be noted. A quick scan of the table of contents reveals **Article 408, Switchboards, Switchgear, and Panelboards**. **Part III** is **Panelboards**. After reading through **Part III**, however, the user will find that there is no mention of a clothes closet within any of the sections.

This is where knowing definitions and understanding what different electrical devices are can be very helpful. A panelboard is defined in **Article 100** as "A single panel or group of panel units designed for assembly in the form of a single panel, including buses and automatic overcurrent devices..." According to this definition, panelboards have overcurrent devices in them. **Section 240.24** in **Article 240**, which covers the location of overcurrent devices, reveals that overcurrent devices cannot be located in the vicinity of easily

Fact

Articles, like the *Code* book as a whole, start generally and become more specific as the article progresses.

ignitable material, such as clothes closets. See **240.24(D)** in the *NEC.*

240.24 Location in or on Premises.

...

(D) Not in Vicinity of Easily Ignitable Material. Overcurrent devices shall not be located in the vicinity of easily ignitable material, such as in clothes closets.

It is useful to remember that articles in the *NEC* start with general topics and move to more specific topics and applications. **Article 410, Luminaires**, is a good example of this.

Article 410 Luminaires, Lampholders, and Lamps

Part I.	**General**
Part II.	**Luminaire Locations**
Part III.	**Provisions at Luminaire Outlet Boxes, Canopies, and Pans**
Part IV.	**Luminaire Supports**
Part V.	**Grounding**
Part VI.	**Wiring of Luminaires**
Part VII.	**Construction of Luminaires**
Part VIII.	**Installation of Lampholders**
Part IX.	**Lamps and Auxiliary Equipment**
Part X.	**Special Provisions for Flush and Recessed Luminaires**
Part XI.	**Construction of Flush and Recessed Luminaires**
Part XII.	**Special Provisions for Electric-Discharge Lighting Systems of 1000 Volts or Less**
Part XIII.	**Special Provisions for Electric-Discharge Lighting Systems of More Than 1000 Volts**
Part XIV.	**Lighting Track**
Part XV.	**Decorative Lighting and Similar Accessories**
Part XVI.	**Special Provisions for Horticultural Lighting Equipment**

The article starts with topics general to most lighting, such as location, supports, and grounding, and moves to more specific equipment and applications, such as lighting track and horticultural lighting equipment.

It is vital not to become discouraged while beginning to master the use of the *National Electrical Code.* In some cases, keywords may not be immediately apparent. They could be disguised within the meaning, use, or definition of a word in the question. An Electrical Worker's ability to use the *Code* will improve with experience in the industry and with the *NEC.*

EXAMPLES OF THE *CODEOLOGY* METHOD

Example questions will be helpful in becoming familiar with the process and the ways to use the *Codeology* method to find answers. In each example, these steps are followed as a guideline:

1. Qualify the question by looking for keywords or topics in the *NEC.*
2. Based on the keywords, work from specific to general and select the appropriate chapter.
3. Using the table of contents, find the relevant article and the relevant part of that article.
4. Within the part of the article, read the items or topics in bold print, looking for headings that match the keyword or topic.
5. If the information is not there or does not answer the question completely, repeat these steps with another keyword(s) in the question.

In some cases, more than five steps may be needed to fully answer the question.

Example 1

At a panel, how must the branch circuits be identified where there is more than one voltage present at a premise?

1. Identify any keywords that may apply.

   Panel, Branch Circuit

2. What stage of the installation is the question asking about—Plan, Build, or Use? Is there a special chapter that would apply?

   This is in the Plan section of the *NEC.*

3. Find the relevant article and part using the Table of Contents.

   According to the Table of Contents in the *NEC,* **Article 210** is in the wiring section and covers branch circuits. **Part I** is titled **General Provisions**. Identifying and labeling conductors is a general topic, as it can apply to all conductors of a system.

4. Find the relevant section by looking at the bold section titles.

   In **Article 210, Part I, Identification of Branch Circuits** is found in **Section 210.5**.

By reading the whole section and sub-sections, the user can find the requirements for the identification of grounded, grounding, and ungrounded conductors.

Example 2

In a health care facility, what wiring methods are permitted for branch circuits fed from the life safety branch to a patient care space?

1. Identify any keywords that may apply.

   Health Care Facility, Branch Circuit, Life Safety Branch, Patient Care Space

2. What stage of the installation is the question asking about—Plan, Build, or Use? Is there a special chapter that would apply?

   This question addresses a special occupancy, so **Chapter 5** is checked first.

3. Find the relevant article and part using the Table of Contents.

   According to the Table of Contents in the *NEC,* **Article 517, Health Care Facilities**, has two parts that may help: **Part II, Wiring Methods**, and **Part III, Essential Electrical System**. It is important to check all parts that could apply, especially when dealing with the special chapters of the *NEC.*

4. Find the relevant section by looking at the bold section titles.

   In **Part II**, **517.12** states that the wiring method must comply with **Chapters 1** through **4**. ***NEC* 517.12(A)** states that whatever method is chosen, it must have the ability to provide an effective ground-fault current path and refers the reader to **Section 250.118**.

5. If the information is not there or does not answer the question completely, repeat these steps with another keyword(s) in the question.

   Article 517, **Part II** revealed general requirements for the raceway for branch circuits serving the patient care space. Another keyword is in the question that needs to be resolved. In this case, knowledge of these systems will play an important part in finding the information. **Part III** of **Article 517** is the next place to look, as indicated by the Table of Contents. **Part III** is titled **Essential Electrical System**. Whenever an Electrical Worker is working in a special occupancy for the first time, reading the appropriate article is good practice.

Within **Part III**, **517.31(C), Wiring Requirements,** is found, and **517.31(C)(3)** specifies certain raceways that must be used to provide mechanical protection. It is important to note here that only by reading the entire section can information on the type of raceways be found.

Also note that the question used the words "wiring methods," which the Electrical Worker may interpret as any method found in **Chapter 3** of the *NEC*.

Example 3

What are the requirements, if any, for persons installing medium-voltage cable?

1. Identify any keywords that may apply.

   Medium-Voltage

2. What stage of the installation is the question asking about—Plan, Build, or Use? Is there a special chapter that would apply?

   This is about installing a wiring method, so it is in the Build section, **Chapter 3**.

3. Find the relevant article and part using the Table of Contents.

   According to the Table of Contents, **Article 311** covers medium voltage conductors and cables. Looking at the parts of **Article 311** in the Table of Contents reveals that **Part II** covers **Installation**.

4. Find the relevant section by looking at the bold section titles.

   Section 311.30, Installation, states that "Type MV cable shall be installed, terminated, and tested by qualified persons."

Article 100 contains the definition of *Qualified Person*.

Example 4

Where can the service for a mobile home be located?

1. Identify any keywords that may apply.

   Service, Mobile Home

2. What stage of the installation is the question asking about—Plan, Build, or Use? Is there a special chapter that would apply?

   Services are in the Plan chapter, but a mobile home is a special occupancy; check **Chapter 5** first.

3. Find the relevant article and part using the Table of Contents.

   According to the Table of Contents in **Chapter 5**, **Article 550** covers mobile homes. **Part III** of this article covers **Services and Feeders**.

4. Find the relevant section by looking at the bold section titles.

   Section 550.32, Service Equipment, states that there are two possible locations, detailed in **550.32(A)** and **550.32(B)**, depending on the circumstances.

Example 5

Are there any special requirements for receptacles around in-ground swimming pools?

1. Identify any keywords that may apply.

   Receptacle, Swimming Pool

2. What stage of the installation is the question asking about—Plan, Build, or Use? Is there a special chapter that would apply?

   Receptacles are in the Use chapter, but swimming pools are special equipment; check **Chapter 6** first.

3. Find the relevant article and part using the Table of Contents.

   According to the Table of Contents in **Chapter 6**, **Article 680** covers swimming pool installations. **Part II, Permanently Installed Pools,** applies in this case.

4. Find the relevant section by looking at the bold section titles.

   Section 680.22 covers **Lighting, Receptacles and Equipment**, and **680.22(A)(1)** contains the requirements for receptacles around swimming pools.

Example 6

How may a conduit body be supported by EMT?

1. Identify any keywords that may apply.

   Conduit Body, EMT, Support

2. What stage of the installation is the question asking about—Plan, Build, or Use? Is there a special chapter that would apply?

   EMT and conduit bodies are used to build an installation, so the Build chapter, **Chapter 3**, should be consulted.

3. Find the relevant article and part using the Table of Contents.

   According to the Table of Contents in **Chapter 3**, **Article 314** includes conduit bodies, and **Part II** covers **Installation**.

4. Find the relevant section by looking at the bold section titles.

   Section 314.23 covers **Supports**, and **314.23(E)** contains the requirements for using raceways to support enclosures. Conduit bodies are not mentioned in this paragraph; however, remember that it is important to read the entire section, including exceptions. The **Exception** to **314.23(E)** allows conduit bodies to be supported by EMT when sized properly.

   Note there are separate support requirements for the EMT; **314.23(E) Exception** only covers the support of the conduit body.

SUMMARY

With the use of the *Codeology* method in combination with industry experience and knowledge, the *NEC* can be used productively to answer questions and find the requirements needed for installations. Each article is written to move from general to specific topics, so broad topics such as location and support are typically found at the beginning of each article. More specific topics, such as those relating to specific types of equipment, tend to be placed later in the article. Knowing how the *NEC* is organized into Plan, Build, Use, and Special Chapters will guide the reader to finding the correct information.

Using **Section 90.3** as an outline and moving from specific to general, the user can identify which chapter is likely to contain the information he or she is looking for. By using the Table of Contents, the user can then locate the relevant part of the article. Once the correct part is found in the article, reading the section titles in bold print can further narrow down the search until the relevant information is found.

QUIZ

Identify the keyword(s) in the following statements or questions. Choose the most appropriate keyword(s) to be used and the order in which they should be looked up to find the correct information in the *NEC*.

1. **A circuit breaker without a marking indicating the interrupting rating is able to interrupt 5,000 amperes.**
   a. Circuit breaker, Indicating
   b. Circuit breaker, Marking
   c. Interrupting, Amperes
   d. None of the above

2. **Do the receptacles serving the kitchen countertop in a mobile home require GFCI protection?**
   a. Mobile Home, Kitchen Countertop
   b. Mobile Home, Receptacle
   c. Protection, Receptacle
   d. Receptacle, Kitchen Countertop

3. **How are communications systems in a mobile home grounded?**
   a. Communications, Mobile Home
   b. Grounding, Communications
   c. Grounding, Mobile Home
   d. Mobile Home, Communications

4. **The ampacity of a motor branch circuit conductor is generally 125% of the full-load amperes of the motor.**
   a. Branch Circuit, Motor
   b. Motor, Ampacity
   c. Motor, Branch Circuit
   d. Motor, Full Load Amperes

5. **What is the required width of working space for all electrical equipment?**
   a. All Electrical Equipment, Working Space
   b. Width, Working Space
   c. Working Space, Equipment
   d. None of the above

6. **What is the ampacity of conductors supplying a phase converter that provides power to variable loads?**
   a. Ampacity, Conductors
   b. Conductors, Phase Converter
   c. Phase Converter, Conductors
   d. Variable Loads, Provide Power

QUIZ

7. **All fire alarm system circuits must be installed to comply with Section 300.4.**
   a. All Fire Alarm Systems, **300.4**
   b. Circuits, **300.4**
   c. Circuits, Fire Alarm Systems
   d. Installation, **300.4**

8. **Wiring of emergency systems must be kept separate from wiring of normal systems.**
   a. Emergency Systems, Separate
   b. Emergency Systems, Wiring
   c. Wiring, Emergency Systems
   d. Wiring, Separate

9. **IMC must be secured within three feet of a junction box.**
   a. IMC, Junction Box
   b. IMC, Secured
   c. Junction Box, Secured
   d. Secured, IMC

Answer the questions below using a copy of the *NEC* and by reviewing Chapter 10 of this text.

10. **Which keyword or topic is not associated with Chapter 6 of the *NEC*?**
    a. Electric vehicle supply equipment
    b. Motor controllers
    c. Office furniture
    d. Pool pump motors

11. **Which chapter of the *NEC* is called the "Build" chapter in *Codeology*?**
    a. **Chapter 1**
    b. **Chapter 2**
    c. **Chapter 3**
    d. **Chapter 4**

12. **Which keyword or topic is associated with Chapter 2 of the *NEC*?**
    a. Raceway
    b. Grounded Conductor
    c. Industrial Equipment
    d. Ampacity

13. **Which keyword or topic is associated with Chapter 7 of the *NEC*?**
    a. Communication
    b. Fixture wire
    c. Flexible cable
    d. Low-voltage wiring

14. **Which keyword or topic is not associated with Chapter 1 of the *NEC*?**
    a. Working Clearance
    b. Tunnel Installations
    c. Definitions
    d. Support of Raceways

15. **When installing MC cable in a dairy barn, what article(s) should be consulted? (Choose more than one)**
    a. **Article 300**
    b. **Article 330**
    c. **Article 334**
    d. **Article 547**

How the *NEC* is Made

NFPA 70: National Electrical Code (NEC) is revised by the National Fire Protection Association (NFPA) every three years and made available for use by designers, code enforcers, contractors, installers, and other *Code* users. Three years prior to a new release of an NFPA document, NFPA Technical Committee members and staff collaborate extensively with task groups to consider changes presented to the Code-Making Panels (CMPs) by *Code* users and the public, and to address other issues identified in the previous change cycle. Special task groups work on technical and usability issues for consideration by the CMPs. The *NEC* is a true work in progress—a dynamic document that is shaped, molded, and improved by all members of the public who take part in the process.

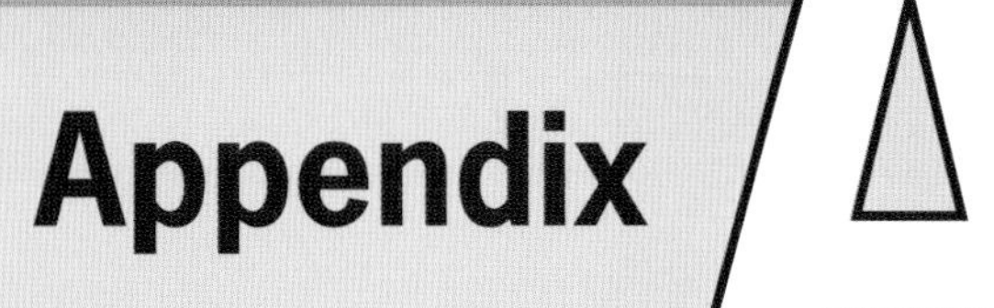

Appendix A

Table of Contents

THE NFPA AND THE *CODE*-MAKING PROCESS

The NFPA was founded in 1896 in Boston by a small group of men who saw the need to set standards for protecting the public against fire hazards. The name has been altered throughout the years, but the initial goal remains the same: protection of the public. Initially, the organization developed recommended practices for fire suppression using sprinklered water and addressed the design and installation of early electrical distribution systems, but over the past 100 years, the NFPA has become the code-making organization for many of the fire protection codes adopted by most U.S. municipalities.

Structure of the NFPA

The NFPA creates minimum codes and standards for fire prevention and other life-safety codes and standards for the United States. The NFPA Board of Directors is the top governing body, presiding over all 385 codes and standards that the NFPA maintains. The Board of Directors appoints a Standard Council of 13 members that oversees and administers the rules and regulations. The Standard Council appoints all CMPs and Technical Committees (TCs), Correlating Committees (CCs), and Motion Committees. In addition, the Standard Council forms new CMPs and TCs if new technologies or situations warrant them. Both the NFPA and *NEC* contain a basic hierarchy of governing bodies. **See Figure A-1.**

Among the NFPA codes and standards are several electrical-related documents, such as *NFPA 70: National Electrical Code*, *NFPA 70E: Standard for Electrical Safety in the Workplace*, and *NFPA 72: National Fire Alarm and Signaling Code.* There are several NFPA codes that the Electrical Worker may be required to review during the layout or installation stages of a project. **See Figure A-2.**

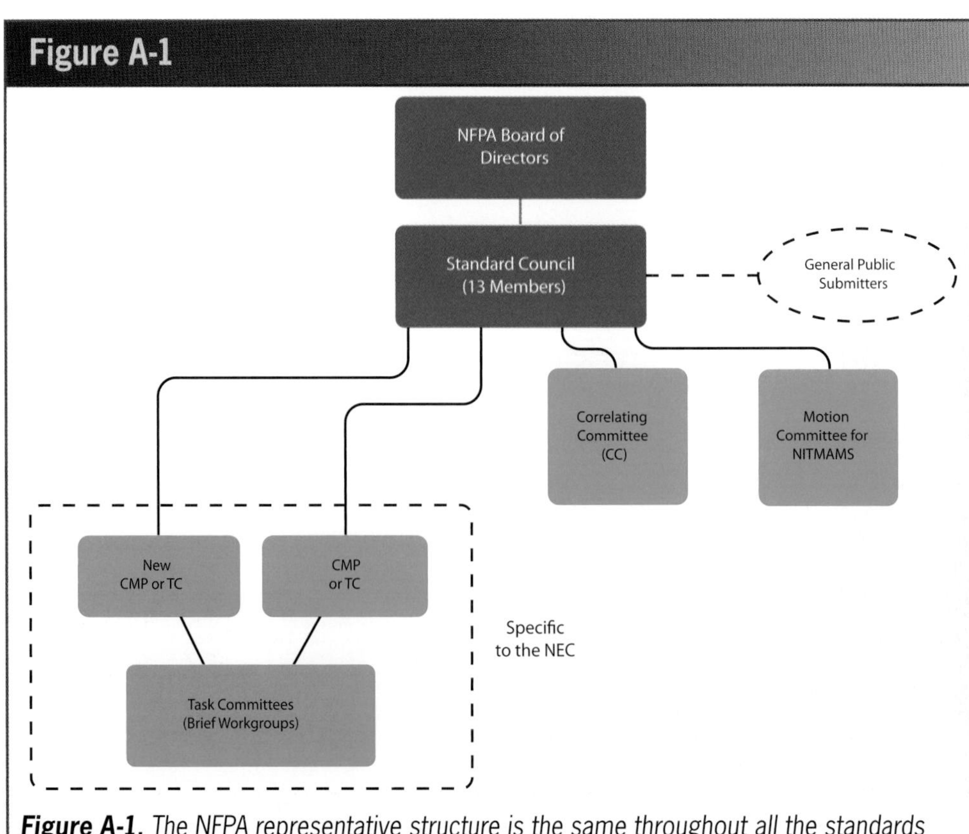

***Figure A-1.** The NFPA representative structure is the same throughout all the standards and codes.*

Figure A-2

NFPA 1	*Fire Code*
NFPA 3	*Recommended Practice on Commissioning and Integrated Testing of Fire Protection and Life Safety Systems*
NFPA 20	*Standard for the Installation of Stationary Pumps for Fire Protection*
NFPA 30A	*Code for Motor Fuel Dispensing Facilities and Repair Garages*
NFPA 70	*National Electrical Code*
NFPA 70A	*National Electrical Code Requirements for One- and Two-Family Dwellings*
NFPA 70B	*Recommended Practice for Electrical Equipment Maintenance*
NFPA 70E	*Standard for Electrical Safety in the Workplace*
NFPA 72	*National Fire Alarm and Signaling Code*
NFPA 73	*Standard for Electrical Inspections of Existing Dwellings*
NFPA 75	*Standard for the Fire Protection of Information Technology Equipment*
NFPA 76	*Standard for the Fire Protection of Telecommunications Facilities*
NFPA 79	*Electrical Standard for Industrial Machinery*
NFPA 92	*Standard for Smoke Control Systems*
NFPA 96	*Standard for Ventilation Control and Fire Protection of Commercial Cooking Operations*
NFPA 99	*Health Care Facilities Code*
NFPA 101	*Life Safety Code*
NFPA 110	*Standard for Emergency and Standby Power Systems*
NFPA 111	*Standard on Stored Electrical Energy Emergency and Standby Power Systems*
NFPA 170	*Standard for Fire Safety and Emergency Symbols*
NFPA 220	*Standard on Types of Building Construction*
NFPA 262	*Standard Method of Test for Flame Travel and Smoke of Wires and Cables for Use in Air-Handling Spaces*
NFPA 450	*Guide for Emergency Medical Services and Systems*
NFPA 469	*Standard for Purged and Pressurized Enclosures for Electrical Equipment*
NFPA 497	*Recommended Practice for the Classification of Flammable Liquids, Gases, or Vapors and of Hazardous [Classified] Locations for Electrical Installations in Chemical Process Areas*
NFPA 499	*Recommended Practice for the Classification of Combustible Dusts and of Hazardous [Classified] Locations for Electrical Installations in Chemical Process Areas*
NFPA 654	*Standard for the Prevention of Fire and Dust Explosions from the Manufacturing, Processing, and Handling of Combustible Particulate Solids*
NFPA 730	*Guide for Premises Security*
NFPA 731	*Standard for the Installation of Electronic Premises Security Systems*
NFPA 780	*Standard for the Installation of Lightning Protection Systems*
NFPA 791	*Recommended Practice and Procedures for Unlabeled Electrical Equipment Evaluation*
NFPA 850	*Recommended Practice for Fire Protection for Electric Generating Plants and High Voltage Direct Current Converter Stations*
NFPA 853	*Standard for the Installation of Stationary Fuel Cell Power Systems*
NFPA 5000	*Building Construction and Safety Code*

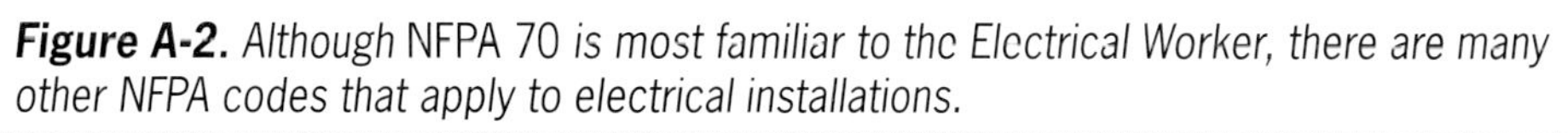
Figure A-2. *Although NFPA 70 is most familiar to the Electrical Worker, there are many other NFPA codes that apply to electrical installations.*

For additional information, visit qr.njatcdb.org Item #1060

Keep in mind that all 385 NFPA codes and committees are revised every three years, but are not on the same three-year schedule. For instance, the *NFPA 70 (NEC)* revision schedule is 2020, 2023, and so on, while the *NFPA 72* revision schedule is 2019, 2022, and so on.

NEC Process

The *NEC* is developed through a consensus standards development process approved by the American National Standards Institute (ANSI). This process includes input from all interested parties (including the general public) throughout the first revision, second revision, and amending-motions stages. CMPs are formed to achieve consensus on all proposed changes or revisions to the *NEC*. These volunteer committees, with members representing various classifications, are balanced to ensure all viewpoints of all interest groups have a voice in deliberating issues brought before the *NEC*.

As defined in the NFPA *Manual of Style for NFPA Technical Committee Documents*, "consensus," as applied to NFPA documents, means that a substantial agreement to a proposed change of language in a code or standard has been reached by the affected interest categories. "Substantial agreement" requires much more than a simple majority; two-thirds of the CMP members must agree, though the vote to adopt a proposal does not necessarily need to be unanimous.

It is important for all users of the *NEC* to understand the document revision process. This process allows for public participation in the development of the *NEC*.

Each revision cycle has an established schedule of steps published in the back of the *Code* book. The process has four basic steps:

1. Input stage
2. Comment stage
3. Association technical meeting
4. Council appeals and issuance of standard

Throughout the revision cycle, the process produces the *First Draft Report* (FDR), previously known as the Report on Proposals (ROP); a *Second Draft Report* (SDR), previously known as Report on Comments (ROC); and Certified Amended Motions for the Association Technical Meeting (NFPA National Meeting). The older method, using ROP and ROC, was last used in the creation of the 2014 edition of the *NEC*. While the ROP and ROC method is no longer used, it is important to be aware of it in case a user wishes to understand a particular code issue and its history. Associated reports of older editions of the *NEC* are available online.

The CMPs (sometimes also called Technical Committees, or TCs) work diligently to provide the *NEC* user with *Code* text that is practical, easy to read and understand, and enforceable. However, as *Code* changes are applied, users may disagree about the implementation of some of the revised requirements. The intent of the changes can be found in the substantiations, CMP statements, and CMP member comments at both the first draft and second draft report stages.

All persons who submit a proposal are notified of the committee's action and have access to the first draft report and second draft report on the NFPA's website. These documents are extremely valuable for the user of the next edition of the *NEC*.

In the back of each edition of the *NEC* are instructions for participating in the *Code* change process. Anyone can submit a proposed change to the next edition of the *NEC*. Most proposals will be submitted online at the NFPA website, www.nfpa.org.

SEQUENCE OF EVENTS

Starting with the 2017 *NEC* cycle, the NFPA moved to a new format for updating the *National Electrical Code*. The following explains the consensus standards development process. Actual deadlines for each of these steps are listed on the NFPA website and in the back of the *NEC*. **See Figure A-3.**

Public Input

One of the first steps in the change process is the public input stage. This process begins when a person wishing to submit a code change accesses the Standards Development Site on the NFPA web page, where there is a tutorial and clear instructions for submitting changes.

Soon after an edition of the *NEC* is published, it is open to public inputs. A public notice requesting interested parties to submit specific written proposals is published in the NFPA News, the U.S. Federal Register, the American National Standards Institute's Standards Action, the NFPA website, and other publications. In this stage, all proposals are submitted electronically on the NFPA website. Proposed changes to the *Code* are submitted to the NFPA, which assigns them to a TC or CC for consideration in the next edition.

The public inputs are created by the interested party and submitted online under the following categories:

- New additions to the *Code*
- Revision of an existing section
- Creation of a global revision to add, modify, or delete a word or phrase throughout the entire document

The last point is a major reason that Electrical Workers need to take the time to make public inputs. Writing code that is understood the same way by anyone who reads it is challenging. If a code section is unclear, submit a public input. Public inputs can aid the members of the Code-Making Panels in creating good, usable, and easy-to-understand code.

During the online input submittal process, once a topic has been selected, the table of contents in the *NEC* appears to help navigate to the pertinent chapter, part, article, and section. All changes are shown in proper editorial format to ensure clarity. The submitter can save, change, or delete a proposal at any time before the public input submission deadline date.

For additional information, visit qr.njatcdb.org Item #1061

Input Stage – First Draft Report

Once the public input date has passed, the proposals are forwarded to the appropriate CMP to be addressed at the Public Input meeting. In this public meeting, the CMP reviews and votes on the proposals and develops the first draft of the *Code*. The CMPs meet at the location and dates stated in the back of the *NEC* to act on all public inputs. These meetings are open to all interested persons who want to observe the panel proceedings. Actions at the meetings require only a simple majority vote. Public inputs that receive a majority vote move on to the ballot.

Figure A-3

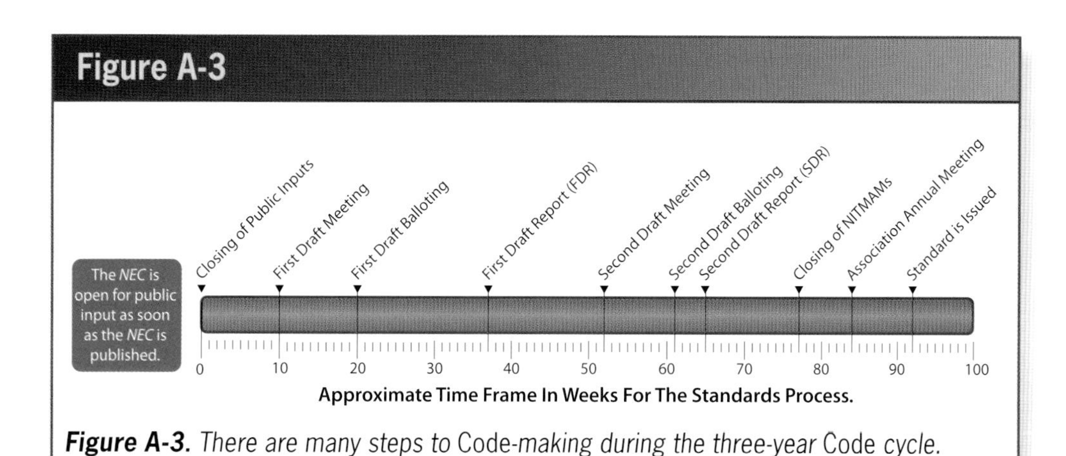

Figure A-3. *There are many steps to Code-making during the three-year Code cycle.*

After the CMP meetings for the first draft have closed, the ballots are sent to all panel members. A two-thirds majority on the ballot is required for the panel action to be upheld. All ballots are filed electronically.

After completion of the public input meeting and balloting, the first draft report (FDR)is issued, and all submitters are notified of the committee's action. The first draft report is available to the public, submitters, and CMPs.

Public Comment Stage and the Second Draft Report

Next is the public comment stage. Once the first draft report becomes available, there is a public comment period during which anyone may submit a public comment on the first draft. Any objections or further related changes to the content of the first draft must be submitted at the comment stage; however, no new material is allowed to be considered.

After the public comment closing date, the committee holds their second draft meeting. All the public comments are considered, and the committee provides an action and response to each public comment. These are no longer called public inputs, but are now called second revisions. The committee will use the public comments in order to help develop second revisions to the standard, resulting in a complete and fully integrated draft known as the second draft report (SDR).

Like the first draft, the second draft has initial agreement by the committee based on a simple majority vote during the meeting to establish a consensus. Next, the second revisions developed at the second draft meeting are balloted. This means that the text the committee wants revised in the standard is on the ballot for approval by the committee. Only those second revisions that garner two-thirds of the balloted vote appear in the second draft. Any second revisions that do not pass the ballot appear in the second draft report as committee comments.

The second draft report is posted on the NFPA website, where it serves as documentation of the comment stage and is published for public review. This second draft for the most part represents what the new edition will look like, unless these second revisions are challenged by amending motions.

NFPA Annual Association Meeting and Amending Motions

The next step in the process is the Annual Association Meeting. Following the completion of the input and comment stages, there is further opportunity for debate and discussion of issues through the Annual Association Meeting (Annual Technical Meeting), which takes place at the NFPA Conference & Expo. The full *NEC* Technical Committee report is presented to the NFPA membership for approval at the NFPA Annual Association Meeting. Because the *NEC* revision is made available to the public for revision years 2020, 2023, and so on, the annual meetings that include the *NEC* amending motions and approvals are held during the intermediate years of 2019, 2022, and so on.

Motions may be made to amend or reverse the actions taken by the CMPs at the annual meeting. Before making a motion at the Annual Association Meeting, the intended maker of the motion must file a *notice of intent to make a motion* (NITMAM) 30 days in advance. The cycle schedule will list the deadline date to submit a NITMAM before the annual meeting. The NITMAM will be received and approved by the Motions Committee and listed on the approved motion listing as a Certified Amending Motion (CAM) for the annual meeting. At the annual meeting, the submitter (or representative) of the NITMAM must notify the NFPA one hour prior to the meeting start time that they will be making the motion. Debate and voting by NFPA members will be allowed at the annual meeting on all CAMs and will carry with a simple majority vote.

Appeals to the Standards Council

An appeal to actions taken at the annual meeting can be made. Appeals to the Standards Council may be made up to 20 days after the annual meeting, after which the Standards Council adjudicates any appeals, accepts the new *NEC*, and issues the revised *Code*.

NEC COMMITTEES

In addition to the Standards Council, the NFPA uses specific committees in the development of the *NEC*. These committees collaborate to ensure that the newly developed code proposals are best suited throughout all the NFPA codes. The committees also perform technical research to better understand the future needs of public safety within the electrical industry. Some of these committees are maintained as part of the standard *Code*-making process, while other committees may be briefly convened to perform a particular task and then dissolved shortly afterward.

Correlating Committee

The Correlating Committee (CC) reviews all of the CMP results to ensure that there are no conflicting actions. Their duties are extensive because they are required to review multiple CMPs' second draft reports. Typically, the chairmen of the CMPs will advise the CC to ensure that possible conflicting actions are resolved. **See Figure A-4.**

Figure A-4

COMMITTEE PERSONNEL

NATIONAL ELECTRICAL CODE COMMITTEE

These lists represent the membership at the time the Committee was balloted on the final text of this edition. Since that time, changes in the membership may have occurred. A key to classifications is found at the back of this document.

Correlating Committee on National Electrical Code®

Michael J. Johnston, *Chair*
National Electrical Contractors Association, MD [IM]

Mark W. Earley, *Secretary (Nonvoting)*
National Fire Protection Association, MA

Sarah D. Caldwell, *Technical Committee Administrator (Nonvoting)*
National Fire Protection Association, MA

James E. Brunssen, Telcordia Technologies (Ericsson), NJ [UT]
Rep. Alliance for Telecommunications Industry Solutions
Kevin L. Dressman, U.S. Department of Energy, MD [U]
Palmer L. Hickman, Electrical Training Alliance, MD [L]
Rep. International Brotherhood of Electrical Workers
David L. Hittinger, Independent Electrical Contractors of Greater Cincinnati, OH [IM]
Rep. Independent Electrical Contractors, Inc.
Richard A. Holub, The DuPont Company, Inc., DE [U]
Rep. American Chemistry Council
John R. Kovacik, UL LLC, IL [RT]
Alan Manche, Schneider Electric, KY [M]
Roger D. McDaniel, Georgia Power Company, GA [UT]
Rep. Electric Light & Power Group/EEI
David A. Williams, Delta Charter Township, MI [E]
Rep. International Association of Electrical Inspectors

Alternates

Lawrence S. Ayer, Biz Com Electric, Inc., OH [IM]
(Alt. to David L. Hittinger)
Roland E. Deike, Jr., CenterPoint Energy, Inc., TX [UT]
(Alt. to Roger D. McDaniel)
James T. Dollard, Jr., IBEW Local Union 98, PA [L]
(Alt. to Palmer L. Hickman)
Ernest J. Gallo, Telcordia Technologies (Ericsson), NJ [UT]
(Alt. to James E. Brunssen)
Robert A. McCullough, Tuckerton, NJ [E]
(Alt. to David A. Williams)
Robert D. Osborne, UL LLC, NC [RT]
(Alt. to John R. Kovacik)
Christine T. Porter, Intertek Testing Services, WA [RT]
(Voting Alt.)
George A. Straniero, AFC Cable Systems, Inc., NJ [M]
(Voting Alt.)

Nonvoting

Timothy J. Pope, Canadian Standards Association, Canada [SE]
Rep. CSA/Canadian Electrical Code Committee
William R. Drake, Fairfield, CA [M]
(Member Emeritus)
D. Harold Ware, Libra Electric Company, OK [IM]
(Member Emeritus)

Mark W. Earley, NFPA Staff Liaison

Figure A-4. *The Correlating Committee is the watchdog that ensures there are no conflicting actions between CMPs.*

Code-Making Panels

Presently, the *NEC* CMPs consist of eighteen panels. As new technologies develop, the Standard Council will recognize the need to derive new CMPs. **See Figure A-5.**

The International Brotherhood of Electrical Workers (IBEW) and the National Electrical Contractors Association (NECA) are represented on many of the NFPA CMPs. All the *NEC* panels and members are listed in the front of the *NEC*.

Figure A-5

NEC Code-Making Panel	Articles, Annex, and Chapter 9 Material Within the Scope of the Code-Making Panel
1	90; 100; 110; Chapter 9, Table 10; Annex A; Annex H; Annex I; Annex J
2	210; 220; Annex D, Examples D1 through D6
3	300; 590; 720; 725; 727; 728; 760; Chapter 9, Tables 11(A) and (B), Tables 12(A) and (B)
4	690; 692; 694; 705; 710
5	200; 250
6	310; 311; 320; 322; 324; 326; 330; 332; 334; 336; 337; 338; 340; 382; 394; 396; 398; 399; 400; Chapter 9, Tables 5 through 9; Annex B, Example D7
7	545; 547; 550; 551; 552; 555; 604; 675; Annex D, Examples D11 and D12
8	342; 344; 348; 350; 352; 353; 354; 355; 356; 359; 360; 362; 366; 369; 370; 372; 374; 376; 378; 380; 384; 386; 388; 390; 392; Chapter 9, Tables 1 through 4; Annex C; Annex D, Example D13
9	312; 314; 404; 408; 450; 490
10	215; 225; 230; 240; 242
11	409; 430; 440; 460; 470; Annex D, Example D8
12	610; 620; 625; 626; 630; 640; 645; 647; 650; 660; 665; 668; 669; 670; 685; Annex D, Examples D9 and D10
13	445; 455; 480; 695; 700; 701; 702; 706; 708; 712; 750; Annex F; Annex G
14	500; 501; 502; 503; 504; 505; 506; 510; 511; 513; 514; 515; 516
15	517; 518; 520; 522; 525; 530; 540
16	770; 800; 805; 810; 820; 830; 840
17	422; 424; 426; 427; 680; 682
18	393; 406; 410; 411; 600; 605

Figure A-5. *There are 18 CMPs for the* NEC.

Likewise, for other NFPA codes and standards (for example, *NFPA 72*), the CMP members are listed in the front of the publication. Reviewing the CMP listing can provide a sense of who is acting as the driving force behind the *Code*-making process. **See Figure A-6.**

COMMITTEE MEMBERSHIP

Members of the NFPA committees are made up of subject-matter experts who are employed or associated with the subject of the committee. As discussed earlier, the committees are made up of volunteer users of the *Code* such as members of research/testing labs, enforcing authorities, insurance agencies, consumers, manufacturers, utilities, and special experts.

The committee membership is made up of highly-qualified personnel with public safety as a high priority. To become a member of a CMP, the applicant is required to complete a form that requests a variety of information such as qualifications, relationship to other members, availability to participate actively, the funding source for the applicant's participation, the background of the applicant's employer, and other information such as which organization the applicant would represent.

Classification

The CMPs are made up of volunteers. A listing of the scope for each CMP and the committee list appear in the front of the *NEC*. In the committee

Figure A-6

Code-Making Panel No. 13

Articles 445, 455, 480, 695, 700, 701, 702, 708, 750, Annex F and Annex G

Linda J. Little, *Chair*
IBEW Local 1 Electricians JATC, MO [L]

Martin D. Adams, Adams Electric, Inc., CO [IM]
Rep. National Electrical Contractors Association
Steve Baldwin, Intertek, CA [RT]
Rep. Intertek Testing Services
Greg J. Ball, Tesla, CA [M]
Krista McDonald Biason, HGA Architects and Engineers, MN [U]
Rep. American Society for Healthcare Engineering
Daniel J. Caron, Bard, Rao + Athanas Consulting Engineers, LLC, MA [SE]
Richard D. Currin, Jr., North Carolina State University, NC [U]
Rep. American Society of Agricultural & Biological Engineers
Neil A. Czarnecki, Reliance Controls Corporation, WI [M]
Rep. National Electrical Manufacturers Association
Steven F. Froemming, City of Franklin, WI [E]
Rep. International Association of Electrical Inspectors
Robert E. Jordan, Alabama Power Company, AL [UT]
Rep. Electric Light & Power Group/EEI
John R. Kovacik, UL LLC, IL [RT]
Greg Marchand, Briggs & Stratton, WI [M]
Rep. Portable Generator Manufacturers' Association
Daniel R. Neeser, Eaton's Bussmann Division, MO [M]
Shawn Paulsen, CSA Group, Canada [RT]
Arnoldo L. Rodriguez, LyondellBasell Industries, TX [U]
Rep. American Chemistry Council
Michael L. Savage, Sr., Marion County Building Safety, FL [E]
Mario C. Spina, Verizon Wireless, OH [U]
Rep. Institute of Electrical & Electronics Engineers, Inc.
Kendall M. Waterman, Draka Cableteq, MA [M]
Rep. Copper Development Association Inc.
James R. White, Shermco Industries, Inc., TX [IM]
Rep. InterNational Electrical Testing Association
Timothy P. Windey, Cummins Power Generation, MN [M]

Alternates

Lawrence S. Ayer, Biz Com Electric, Inc., OH [IM]
(Voting Alt.)
Barry S. Bauman, Alliant Energy, WI [U]
(Alt. to Richard D. Currin, Jr.)
Glenn Brown, CJL Engineering, PA [U]
(Alt. to Krista McDonald Biason)
William P. Cantor, TPI Corporation, PA [U]
(Alt. to Mario C. Spina)
James S. Conrad, RSCC Wire & Cable, CT [M]
(Alt. to Kendall M. Waterman)
Timothy Crnko, Eaton's Bussmann Business, MO [M]
(Alt. to Daniel R. Neeser)
James T. Dollard, Jr., IBEW Local Union 98, PA [L]
(Alt. to Linda J. Little)
Laurie B. Florence, UL LLC, IL [RT]
(Alt. to John R. Kovacik)
Travis Foster, Shell Oil Company, TX [U]
(Alt. to Arnoldo L. Rodriguez)
Richard Garbark, BGE, MD [UT]
(Alt. to Robert E. Jordan)
Jan Gromadzki, Tesla, CA [M]
(Alt. to Greg J. Ball)
Jeff Jonas, Generac Power Systems, Inc., WI [M]
(Voting Alt.)
Chad Kennedy, Schneider Electric, SC [M]
(Alt. to Neil A. Czarnecki)
Raymond Richard Prucha, Bard, Rao + Anthanas Consulting Engineers, LLC, NY [SE]
(Alt. to Daniel J. Caron)
Daniel Schlepp, Wacker Neuson, WI [M]
(Alt. to Greg Marchand)
Rich Scroggins, Cummins Power Generation, MN [M]
(Alt. to Timothy P. Windey)
Richard Tice, VEC, Inc., OH [IM]
(Alt. to Martin D. Adams)
Anton Tomasin, City Of Rochester Hills, MI [E]
(Alt. to Steven F. Froemming)
Michael Wilson, CSA Group, Canada [RT]
(Alt. to Shawn Paulsen)

Figure A-6. *CMP No. 13 for the NEC has an IBEW chair from St. Louis, Missouri.*

list, CMP member names are followed by their employer names and identification letter(s). The identification letters appear in brackets, such as [L] for labor. Most organizations represented have a principal member and an alternate member.

Committee membership classification is part of the balancing process of each panel. **See Figure A-7.**

Representation

Many organizations take part in the *NEC* process and provide representation for the membership classification that applies to their organization. **See Figures A-8 and A-9.** For example, the members of the CMPs with the classification [L] for labor are representatives of the International Brotherhood of Electrical Workers (IBEW).

Figure A-7

M	Manufacturers: makers of products affected by the *NEC*
U	Users: users of the *NEC*
I/M	Installers/Maintainers: installers/maintainers of systems covered by the *NEC*
L	Labor: those concerned with safety in the workplace
R/T	Research/Testing Labs: independent organizations developing/reinforcing standards
E	Enforcing Authority: inspectors and other enforcers of the *NEC*
I	Insurance: insurance companies, bureaus, or agencies
C	Consumers: purchasers of products/systems not included in "U" (users)
SE	Special Experts: providers of special expertise, not applicable to other classifications
UT	Utilities: installers/maintainers of systems not covered by the *NEC*

Figure A-7. *CMP members come from a wide background of expertise, such as manufacturers and research labs.*

Figure A-8

NFPA Electrical Engineering Division Technical Staff

Gordon S. Frost, Division Manager
Mark W. Earley, Chief Electrical Engineer
Sarah D. Caldwell, Technical Committee Administrator
Kimberly H. Cervantes, Senior Technical Content Editor
Barry Chase, Senior Engineer
Mark Cloutier, Senior Electrical Engineer
Christopher Coache, Senior Electrical Engineer
Erik Hohengasser, Technical Lead, Electrical
Richard J. Roux, Senior Electrical Specialist
Jeffrey S. Sargent, Principal Electrical Specialist

Figure A-8. *The NFPA staff consist of technical, support, and editorial personnel.*

Figure A-9

Air-Conditioning, Heating, & Refrigeration Institute	Instrumentation, Systems, & Automation Society
Alliance for Telecommunications Industry Solutions	Insulated Cable Engineers Association Incorporated
Alliance of Motion Picture and Television Producers	International Association of Electrical Inspectors
The Aluminum Association Incorporated	International Alliance of Theatrical Stage Employees
American Chemistry Council	International Electrical Testing Association Incorporated
American Iron and Steel Institute	International Brotherhood of Electrical Workers
American Institute of Organ Builders	International Sign Association
American Lighting Association	National Association of Homebuilders
American Petroleum Institute	National Association of RV Parks and Campgrounds
American Society of Agricultural & Biological Engineers	National Cable & Telecommunications Association
American Society for Healthcare Engineering	National Electrical Contractors Association
American Wind Energy Association	National Electrical Manufacturers Association
Associated Builders and Contractors	National Elevator Industry Incorporated
Association of Higher Education Facilities Officers	Outdoor Amusement Business Association Incorporated
Association of Pool & Spa Professionals	Power Tool Institute Incorporated
Automatic Fire Alarm Association	Recreational Vehicle Industry Association
Building Industry Consulting Service International	Satellite Broadcasting & Communications Association
CSA/Canadian Electrical *Code* Committee	Society of Automotive Engineers – Hybrid Committee
Copper Development Association Incorporated	Society of the Plastics Industry Incorporated
Electric Light & Power Group/EEI	Solar Energy Industries Association
Electronic Security Association	TC on Airport Facilities
Grain Elevator and Processing Safety	TC on Electrical Systems
Illuminating Engineering Society of North America	Telecommunications Industry Association
Independent Electrical Contractors	Transportation Electrification Committee
Information Technology Industry Council	U.S. Institute for Theatre Technology
Institute of Electrical & Electronics Engineers	The Vinyl Institute

Figure A-9. *Many interested organizations are members of CMPs.*

SUMMARY

The *NEC* revision process is open to all members of the public who wish to take part by submitting proposed changes. The many organizations that participate in the *NEC Code*-making process help to build a true consensus code.

Since the *NEC* will change every three years, attempting to memorize requirements or sections may be a wasted effort. The *Codeology* method, however, will not change. Applying the *Codeology* method will lead to the quick and accurate location of necessary information in the *NEC* today and in all future and past versions of the *Code.*

Code Tests

This section of the textbook contains two code tests that cover information from each chapter of the *NEC*. Each test includes 50 questions and provides the user with opportunities to practice applying the *Codeology* method. In each question, choose the answer that best completes the statement or answers the question. The correct answers and *NEC* sources are located at the end of the tests.

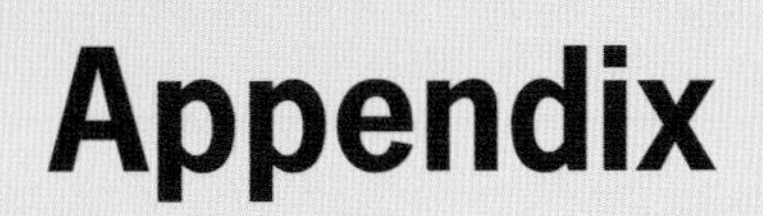

Appendix B

Table of Contents

CODE TEST 1

1. **The branch-circuit selection current of air-conditioning equipment is __?__ the rated load current.**
   a. always equal to
   b. always greater than
   c. equal to or greater than
   d. less than

2. **For dwelling unit services of 100 amperes, the service conductors shall be permitted to have an ampacity as low as __?__.**
   a. 63 A
   b. 83 A
   c. 90 A
   d. 100 A

3. **Which of the following materials are permitted to construct RMC?**
   a. Brass
   b. Copper
   c. Red brass
   d. Steel, without protective covering

4. **A bathroom must have which of the following?**
   a. Bidet
   b. Sink
   c. Toilet
   d. Urinal

5. **ITC cable may not be used on circuits operating at more than __?__ to ground.**
   a. 60 V
   b. 120 V
   c. 150 V
   d. 300 V

6. **The ampacity for a 60°C rated 24 AWG conductor used for control wiring on a permanent amusement attraction is __?__.**
   a. 1 A
   b. 1.19 A
   c. 2 A
   d. 3 A

7. **What is the conduit fill percentage for four conductors in an EMT nipple 36 inches long?**
   a. 31%
   b. 40%
   c. 53%
   d. 60%

8. **Fiber optic cable installations must comply with __?__.**
   a. Article 300
   b. Section 300.4
   c. Section 300.11
   d. both Sections 300.4 and 300.11

9. **The continuity of a grounded conductor shall not depend on __?__.**
   a. a metallic enclosure
   b. cable armor
   c. raceway
   d. any of the above

10. **When connecting an SPD to the line conductors and ground, the installer shall __?__.**
    a. avoid any unnecessary bends
    b. include a service loop
    c. not use XHHW conductors
    d. use XHHW conductors

11. **Refrigerant machinery rooms that contain ammonia-based systems and are continuously ventilated are permitted to be a(n) __?__ location.**
    a. Class 2, Division 1
    b. Class 2, Division 2
    c. Class 3
    d. unclassified

12. **Lighting equipment listed for horticultural use and connected with flexible cord shall have outlets that are __?__.**
    a. GFCI protected
    b. installed 6'7" or higher from the floor
    c. installed not less than 18" from the ceiling
    d. none of the above

13. **When calculating loads, calculations __?__.**
    a. are permitted to be rounded up if the decimal is 0.5 or greater
    b. are permitted to be rounded up in all cases
    c. must be rounded down to the nearest whole ampere
    d. must be rounded up

14. **Which of the following is an example of a soft conversion?**
    a. 2.54" = 1 cm
    b. 25.4 mm = 1"
    c. 26 mm = 1"
    d. 30 mm = 1"

15. The maximum number of 14 AWG RHH conductors permitted in ½-inch EMT is __?__.

a. 3
b. 4
c. 6
d. 7

16. The manufactured phase of a phase converter shall not be connected to a __?__.

a. single-phase load
b. 2-phase load
c. 3-phase load
d. 3-phase motor

17. Generally, the minimum size of conductors for circuits up to 2,000 volts shall be __?__.

a. 18 AWG copper
b. 16 AWG copper
c. 14 AWG copper
d. 12 AWG copper

18. Type AC cable shall be supported every __?__.

a. 3'
b. 3.5'
c. 4'
d. 4.5'

19. The insulation of 14 AWG RHH conductor has a thickness of __?__.

a. 0.45 mm
b. 1.14 mm
c. 1.52 mm
d. none of the above

20. Type 1 surge protective devices are for circuits __?__.

a. 1,000 volts or less
b. 1,000 volts or greater
c. greater than 1,000 volts
d. on ungrounded systems

21. Which article is not modified by Article 800?

a. 805
b. 810
c. 820
d. 830

22. The recommended tightening torque of a screw with 1/8-inch recessed square drive is __?__.

a. 3.2 N-m
b. 11.3 N-m
c. 45 lbf-in
d. none of the above

23. Generally, there must not be more than how many services to a building?

a. 1
b. 2
c. 3
d. 4

24. Which of the following is an acceptable method for connecting an equipment grounding conductor to a piece of equipment?

a. Machine screw secured with a nut
b. Self-tapping green screw
c. Terminal bar
d. Both a. and c.

25. Where nonmetallic-sheathed cable is installed in a raceway and enters a box, the cable assembly __?__.

a. is not required to be secured
b. must be secured within the box
c. must be secured within 12" of the box
d. none of the above; nonmetallic-sheathed cable may never be installed in a raceway

26. Receptacles with a USB charger must __?__.

a. be listed
b. have the Class 2 circuitry external to the receptacle
c. have the Class 2 circuitry internal to the receptacle
d. both a. and c.

27. Where lighting accessories are provided for office furniture and are cord-connected, the cord shall not be longer than __?__.

a. 1'
b. 3'
c. 6'
d. 9'

28. All fire alarm system installations shall comply with __?__.

a. 300.4
b. 358.22
c. 725.41
d. 760.41

29. The particles of combustible dust are __?__ or smaller.

a. 250 microns
b. 500 microns
c. 625 microns
d. 750 microns

30. When a fire pump motor starts, the voltage at the line terminals of the controller shall not be more than __?__ below normal voltage.

a. 3%
b. 5%
c. 10%
d. 15%

31. The branch-circuit conductors for a 100-gallon storage-type water heater rated at 26 amperes shall have an ampacity of at least __?__.

a. 20.8 A
b. 26 A
c. 30 A
d. 32.5 A

32. What Class I group does hydrogen belong to?

a. Group A
b. Group B
c. Group E
d. Group H

33. Assembly occupancies are intended for groups of people of __?__ or more.

a. 50
b. 100
c. 500
d. 1,000

34. For services not rated over 1,000 volts, conductors shall not have a clearance of less than __?__ over the surface of a roof.

a. 4'
b. 6'
c. 8'
d. 10'

35. The neutral of a 120/208-volt feeder shall not be smaller than specified in __?__.

a. Table 250.66
b. Table 250.102(C)
c. Table 250.122
d. Table 310.16

36. Which of the following is not a standard fuse size?

a. 1 A
b. 15 A
c. 30 A
d. None of the above

37. What is the cross-sectional area of strut-type raceway that has a channel size of 1½ inches by three inches?

a. 0.964 in^2
b. 1.542 in^2
c. 3.854 in^2
d. 2,487 mm^2

38. The overall length of the output cable to an electric vehicle generally shall not exceed __?__ in length.

a. 10'
b. 15'
c. 20'
d. 25'

39. Raceway fill for fire-resistive cable systems shall __?__.

a. comply with the listing requirements of the system
b. not apply
c. not be greater than the fill requirements in Table 1 of Chapter 9
d. both a. and c.

40. Disconnects for rides at a fair shall be __?__.

a. located within 6' of the ride
b. readily accessible
c. within sight
d. all of the above

41. A 6 AWG copper conductor run vertically for 250 feet requires __?__ conductor supports.

a. 2
b. 3
c. 4
d. none of the above

42. **A 150 kVA dry-type transformer installed indoors shall be installed in a transformer room with a __?__ fire rating.**

a. 15-minute
b. 30-minute
c. 1-hour
d. 2-hour

43. **Receptacle outlets at a picnic area by a lake must have their GFCI protective devices located at least __?__ above the electrical datum plane.**

a. 6"
b. 12"
c. 24"
d. 36"

44. **The cross-sectional area of a solid 12 AWG aluminum conductor is __?__.**

a. 4.25 mm^2
b. 5.3 mm^2
c. 4,110 circular mils
d. 6,530 circular mils

45. **Flat conductor cable shall be covered with __?__.**

a. accessible laminate flooring
b. carpet squares measuring 1 m or less
c. linoleum squares measuring 12" or less
d. any of the above

46. **A conductor that is encased within a material of composition or thickness that is not recognized by the *NEC* is considered a(n) __?__ conductor.**

a. bare
b. covered
c. energized
d. insulated

47. **Generally, when conductors are installed in parallel, they must be size __?__ or larger.**

a. 1 AWG
b. 1/0 AWG
c. 2/0 AWG
d. 3/0 AWG

48. **The interior of a raceway installed in a wet location is considered to be a __?__ location.**

a. damp
b. dry
c. protected
d. wet

49. **Once a park trailer is set up, it is limited to a gross trailer area of __?__.**

a. 200 ft^2
b. 400 ft^2
c. 600 ft^2
d. 800 ft^2

50. **The maximum circuit current of a photovoltaic module with a short-circuit current rating of nine amperes is __?__.**

a. 7.2 A
b. 11.25 A
c. 13.5 A
d. 14 A

ANSWERS FOR CODE TEST 1

1. c.

   Source: 440.2, definition of "Branch Circuit Selection Current"

2. b.

   Source: 310.12(A)

3. c.

   Source: 344.100

4. b.

   Source: Article 100, Part I, definition of "Bathroom"

5. c.

   Source: 727.5

6. c.

   Source: 522.22

7. b.

   Source: Table 1, Chapter 9

8. d.

   Source: 770.24

9. d.

   Source: 200.2(B)

10. a.

    Source: 242.24

11. d.

    Source: 500.5(A)

12. a.

    Source: 410.184

13. a.

    Source: 220.5(B)

14. b.

    Source: 90.9, Informational Note

15. b.

    Source: Chapter 9, Table 1, Note 1; Annex C, Table C.1

16. a.

    Source: 455.9

17. c.

    Source: 310.3(A)

18. d.

    Source: 320.30(C)

19. b.

    Source: 310.4, Table 310.4(A)

20. a.

    Source: 242, Part II, Informational Note; 242.6

21. b.

    Source: 800.1

22. c.

    Source: Annex I, Table I.3

23. a.

    Source: 230.2

24. d.

    Source: 250.8

25. a.

    Source: 314.17(C)(3)

26. d.

    Source: 406.4

27. d.

    Source: 605.6(B)

28. a.

    Source: 760.24(A)

29. b.

    Source: Article 100, Part III, definition of "Combustible Dust"

30. d.

    Source: 695.7(A)

31. d.

    Source: 422.13

    Note: 125% of 26 = 32.5

32. b.

    Source: 500.6(A)(2)

33. b.

    Source: 518.1

34. c.

    Source: 230.24(A)

35. c.

    Source: 215.2(A)(2)

36. a.

    Source: 240.6

37. c.

    Source: 384.22, Table 384.22

38. d.

    Source: 624.17(B)

39. **d.**

Source: 728.5(C)

40. **d.**

Source: 525.21(A)

41. **b.**

Source: 300.19, Table 300.19

42. **c.**

Source: 450.21(B)

43. **b.**

Source: 682.15(A)

44. **d.**

Source: Chapter 9, Table 8

45. **b.**

Source: 324.41

46. **b.**

Source: Article 100, Part I, definition of "Conductor, Covered"

47. **b.**

Source: 310.10(G)

48. **d.**

Source: 300.9

49. **b.**

Source: 552.2, definition of "Park Trailer"

50. **b.**

Source: 690.8(A)(1)(a)(1)

CODE TEST 2

1. **A compound that is used to prevent passage of hazardous vapor shall have a melting point not less than __?__.**
   a. 100°C
   b. 150°C
   c. 200°F
   d. 250°F

2. **Which of the following optical fiber cable types are allowed to be installed in a fireproof shaft riser?**
   a. OFCR
   b. OFNG
   c. OFNP
   d. Any of the above

3. **The area within five feet of aircraft shall be classified as __?__.**
   a. Class 1, Division 1
   b. Class 1, Division 2
   c. Class 1, Zone 2
   d. either b. or c.

4. **A baseboard heater nameplate shall include which of the following information?**
   a. Amperes
   b. Volts
   c. Watts
   d. All of the above

5. **Type PTF wire has a maximum operating temperature of __?__.**
   a. 60°C
   b. 90°C
   c. 170°C
   d. 250°C

6. **Which area of a dwelling unit does not require AFCI-protected branch circuits?**
   a. Bathrooms
   b. Closets
   c. Kitchen
   d. Sunroom

7. **A 10-kilowatt range must have a minimum branch circuit rating of __?__.**
   a. 30 A
   b. 40 A
   c. 50 A
   d. none of the above; there is no minimum branch circuit rating required

8. **The minimum size of aluminum conductor for a 5,000-volt cable is __?__.**
   a. 8 AWG
   b. 6 AWG
   c. 4 AWG
   d. 3 AWG

9. **The smallest trade size of LFNC that may ever be used is __?__.**
   a. 3/8"
   b. 1/2"
   c. 3/4"
   d. 1"

10. **A premises-powered broadband optical fiber cable buried in a residential lawn shall have a minimum cover of __?__.**
   a. 6"
   b. 12"
   c. 18"
   d. 24"

11. **The maximum rating of an overcurrent protective device to protect 12 AWG aluminum conductor is __?__.**
   a. 10 A
   b. 15 A
   c. 20 A
   d. 25 A

12. **Type PVC conduit that is ½-inch trade size must be supported every __?__.**
   a. 3'
   b. 5'
   c. 6'
   d. 10'

13. **Overload protection for a single-phase motor requires how many overload units?**
   a. 1
   b. 2
   c. 3
   d. None

14. **A single-phase transformer with a 240-volt, five-ampere primary may be protected using primary-only protection sized as high as __?__ of the primary current.**
   a. 100%
   b. 125%
   c. 133%
   d. 167%

15. **How many receptacles are required at a patient bed location in a general care space?**
   a. 8
   b. 11
   c. 14
   d. 16

16. **How many cubic inches is a standard plastic three-inch by two-inch by two-inch device box?**
   a. 10 in^3
   b. 12 in^3
   c. 14 in^3
   d. The value stamped on the box by the manufacturer

17. **When a 6 AWG conductor leaves a terminal from a breaker, how many inches must there be between the breaker terminals and the opposite wall of the cabinet?**
   a. 1.5"
   b. 2"
   c. 3"
   d. 4"

18. **The resistance-type immersion heating elements of an ASME-rated boiler may not be protected by an overcurrent device greater than __?__.**
   a. 90 A
   b. 120 A
   c. 150 A
   d. 175 A

19. **The calculated motor load of a crane motor shall be __?__.**
   a. 100% of the motor nameplate
   b. 125% of the motor nameplate
   c. 125% of the value in Tables 430.247 through 430.250
   d. none of the above

20. **An 8 AWG THW conductor is used to supply an arc welder. The maximum rating of an overcurrent protective device with 75°C rated terminations is __?__.**
   a. 50 A
   b. 70 A
   c. 100 A
   d. 125 A

21. **Positive polarity conductors of a DC microgrid shall be __?__.**
   a. black
   b. blue
   c. red
   d. any of the above are acceptable

22. **Heating cables to prevent the freezing of water pipes are installed. The heating cables shall __?__.**
   a. be secured by the insulation
   b. have GFCI protection
   c. have GFPE protection
   d. none of the above

23. **Which cable can be substituted for Class 3 riser cable (CL3R)?**
   a. CL3P
   b. CMP
   c. CMR
   d. All of the above

24. **Where run vertically, metal wireway shall be supported every __?__.**
   a. 5'
   b. 10'
   c. 15'
   d. 20'

25. **An 8 AWG copper conductor rated 75°C can continuously carry __?__.**
   a. 30 A
   b. 40 A
   c. 50 A
   d. 60 A

26. **The minimum conductor size for Type P cable is __?__.**
   a. 18 AWG
   b. 16 AWG
   c. 14 AWG
   d. 12 AWG

27. **GFCI protection is required for boat hoist outlets where the voltage does not exceed __?__.**
   a. 120 V
   b. 208 V
   c. 240 V
   d. 277 V

28. Which of the following is permitted to be used as a disconnect for a motor?

a. Molded case switch
b. Molded case circuit breaker
c. Motor circuit switch
d. Any of the above

29. The maximum ampacity of an 8 AWG 75°C SCE cable is __?__.

a. 40 A
b. 50 A
c. 55 A
d. 60 A

30. Fixed ladders used to enter manholes must be __?__.

a. aluminum
b. corrosion-resistant
c. nonmetallic
d. stainless steel

31. Industrial x-ray equipment that is designed to be carried by hand is __?__.

a. mobile
b. movable
c. portable
d. transportable

32. When installing a feeder to a relocatable structure, there shall be __?__ unless a listed cable assembly is used.

a. 2 insulated and 1 uninsulated conductor
b. 3 insulated conductors
c. 4 insulated conductors
d. any number of color-coded conductors may be used

33. When a switch is in the off position, all __?__ must be disconnected.

a. conductors, including equipment grounding conductors,
b. grounded conductors
c. ungrounded and grounded conductors
d. ungrounded conductors

34. If the rated current of a capacitor is 17 amperes, what is the minimum ampacity of the conductor supplying the capacitor?

a. 17 A
b. 20 A
c. 21.25 A
d. 22.95 A

35. The phase arrangement of the busses of a panelboard shall be A, B, C, __?__.

a. front to back
b. left to right
c. top to bottom
d. any of the above

36. A Group IIC atmosphere contains __?__.

a. acetylene
b. alcohol
c. methane
d. propane

37. An RV park with 53 sites that have electrical supply shall have __?__ sites equipped with a 30-ampere, 125-volt receptacle.

a. 15
b. 16
c. 37
d. 38

38. The demand factor for a feeder supplying electrified truck parking spaces in USDA hardiness zone 4a is __?__.

a. 45%
b. 55%
c. 57%
d. 63%

39. The minimum size of the grounded conductor for a service supplied by three 600 kcmil copper conductors per phase, where all conductors are in the same raceway, is __?__.

a. 1/0 AWG
b. 2/0 AWG
c. 4/0 AWG
d. 250 kcmil

40. The minimum depth of clear working space in front of high-voltage equipment fed by 4,500 volts is __?__.

a. 4'
b. 5'
c. 6'
d. 8'

41. The maximum voltage drop allowed on a branch circuit supplying a receptacle used to supply sensitive electronic equipment, including the cord supplying the equipment, is __?__.

a. 1%
b. 1.5%
c. 2.5%
d. 3%

42. Where practical, CATV conductors shall be separated from lightning conductors by at least __?__.

a. 3'
b. 4.5'
c. 6'
d. 8'

43. When using FMC to interconnect control panels in an elevator room, the FMC is limited to lengths of __?__ or less.

a. 3'
b. 4'
c. 6'
d. 10'

44. The wire-type equipment grounding conductor is __?__.

a. not required to be larger than the conductors supplying the equipment
b. not required to be larger than the size determined by the overcurrent protective device
c. required to be at least one size smaller than the ungrounded conductor
d. required to be larger than the grounded conductor

45. The adjustment factor for conductor ampacity when seven current-carrying conductors are in the same raceway is __?__.

a. 50%
b. 60%
c. 70%
d. 80%

46. The general lighting load used for calculating the service demand for a convention center is __?__.

a. 1.4 VA/ft^2
b. 1.5 VA/ft^2
c. 1.6 VA/ft^2
d. 15 VA/ft^2

47. The voltage to ground on a 480-volt ungrounded system is __?__.

a. 0 V
b. 120 V
c. 240 V
d. 480 V

48. The enclosure that houses the alternator of a wind turbine is called a __?__.

a. boite
b. cabinet
c. coffer
d. nacelle

49. Junction boxes used for fire alarm conductors shall be painted __?__.

a. blue
b. orange
c. red
d. none of the above

50. Audio equipment supplied by a branch circuit for power shall not be within __?__ of the inside wall of a hot tub.

a. 4.5'
b. 5'
c. 6'
d. 8.5'

ANSWERS FOR CODE TEST 2

1. c.

Source: 500.15(C)(2)

2. d.

Source: 770.154, Table 770.154(a)

3. d.

Source: 513.3(C)(1)

4. b.

Source: 422.60(A)

Note: Of the three items listed, only volts is always required.

5. d.

Source: 402.3, Table 402.3

6. a.

Source: 210.12(A)

7. b.

Source: 210.19(A)(3)

8. b.

Source: 311.12(A), Table 311.12(A)

9. a.

Source: 356.20

10. a.

Source: 840.47(A)(3)

11. b.

Source: 240.4(D)(4)

12. a.

Source: 352.30, Table 352.30

13. a.

Source: 430.37, Table 430.37

14. d.

Source: 430.3, Table 430.3(B)

15. a.

Source: 517.18(B)(1)

16. d.

Source: 314.16(A)(2)

17. b.

Source: 312.6(B)2, Table 312.6(B)

18. c.

Source: 424.72(A)

19. a.

Source: 610.14(E)(1)

20. c.

Source: 630.12(B)

21. c.

Source: 712.25

22. c.

Source: 427.22

23. d.

Source: 725.154

24. c.

Source: 376.30(B)

25. c.

Source: 310.16, Table 310.16; Article 100, Part I, definition of "Ampacity"

26. a.

Source: 337.104

27. c.

Source: 555.9

28. d.

Source: 430.109

29. d.

Source: 400.5, Table 400.5(A)(2)

30. b.

Source: 110.79

31. c.

Source: 660.2

32. c.

Source: 544.22(A)

33. d.

Source: 404.20(B)

34. d.

Source: 460.8(A)

35. d.

Source: 408.3(E)(1)

36. a.

Source: 505.6(A)

37. d.

Source: 551.71(B)

38. b.

Source: 626.11(B), Table 626.11(B)

39. d.

Source: 250.24(C)1, Table 250.102(C)(1)

40. a.

Source: 110.34(A), Table 110.34(A)

41. b.

Source: 647.4(D)(2), Informational Note

42. b.

Source: 800.53

43. c.

Source: 620.21(A)(3)

44. a.

Source: 250.122(A)

45. c.

Source: 310.15(C)(1), Table 310.15(C)(1)

46. a.

Source: 210.12(A), Table 210.12

47. d.

Source: Article 100, Part I, definition of "Voltage to Ground"

48. d.

Source: 694.2

49. d.

Source: 760.24

50. b.

Source: 640.10(A)

A

Index

B

C

O

P

R